Leandro Cosme Oliveira Couto

Geoecological Modelling of the Landscape of Corinto and Diamantina, MG

Leandro Cosme Oliveira Couto

Geoecological Modelling of the Landscape of Corinto and Diamantina, MG

Algebra Maps for Morphodynamics, natural vulnerability to soil loss, natural and environmental vulnerabilities

ScienciaScripts

Imprint
Any brand names and product names mentioned in this book are subject to trademark, brand or patent protection and are trademarks or registered trademarks of their respective holders. The use of brand names, product names, common names, trade names, product descriptions etc. even without a particular marking in this work is in no way to be construed to mean that such names may be regarded as unrestricted in respect of trademark and brand protection legislation and could thus be used by anyone.

Cover image: www.ingimage.com

This book is a translation from the original published under ISBN 978-3-330-99792-9.

Publisher:
Sciencia Scripts
is a trademark of
Dodo Books Indian Ocean Ltd. and OmniScriptum S.R.L publishing group

120 High Road, East Finchley, London, N2 9ED, United Kingdom
Str. Armeneasca 28/1, office 1, Chisinau MD-2012, Republic of Moldova, Europe
Managing Directors: Ieva Konstantinova, Victoria Ursu
info@omniscriptum.com

Printed at: see last page
ISBN: 978-620-3-29058-5

ACKNOWLEDGEMENTS

Full gratitude:

To my parents, my brother Vanilson and to God.

To Professor Dr Luiz Eduardo Panisset Travassos (Jedi Master), whose excellent guidance unfolded in indications, corrections, support and precise conversations, allowing me to become more confident and prepared for experiences in the academic environment.

To Professors João Henrique Rettore Totaro and Alecir Antônio Maciel Moreira, who qualified this dissertation and raised it to a better quality for the Defence.

To the Fundação de Coordenação de Aperfeiçoamento de Pessoal de Nível Superior (Capes), for the fundamental funding of the scholarship.

To Roselane Guimarães for willingly sharing the data she had consolidated during her Master's degree.

To Bruno Durão and Professor Panisset for carrying out the final fieldwork, which gave us a panoramic view of the study area.

SUMMARY

This dissertation seeks to apply three types of geological modelling to the landscape at the contact between the Rio das Velhas plain and the Serra do Espinhaço Meridional. This area appears to be a representative sample of broad geoecological diversity due to its location. The research was based on the concepts of landscape and geosystems, guiding the field work, and allowed the application of morphodynamic modelling, natural vulnerability to soil loss and natural and environmental vulnerabilities. Digital data provided by official organisations was collected and consolidated, allowing three transects to be made and four geosystems to be identified in the study area. From west to east, from the municipality of Corinto to Diamantina, there is the Rio das Velhas Plain, the Monjolos karst, the Serra do Cabral and the Serra do Espinhaço Meridional, the latter comprising three geofacies (West Face, Interfluve and East Face). These geosystems confirm the geodiversity that exists in the contact between the Rio das Velhas plain, where the Monjolos karst is the most fragile geosystem and susceptible to impacts, and the Serra do Espinhaço Meridional, whose physical attributes are resistant but whose vegetation cover is a determining factor in environmental fragility.

Keywords: Landscape; geosystem; karst; Serra do Espinhaço Meridional.

CHAPTER 1

INTRODUCTION

"I say: the real is neither at the exit nor at the arrival: it becomes available to us in the middle of the journey."
(J. Guimarães Rosa, O Grande Sertão: Veredas.)

The theme chosen for this dissertation is a geoecological analysis aimed at a systemic understanding of the landscape circumscribed by the boundaries of Topographic Maps SE-23-Z-A-II (Corinto, MG) and SE-23-Z-A-III (Diamantina, MG). The landscape of these Topographic Maps is located in the Southern Espinhaço and the Rio das Velhas Plain, corresponding to a sample of the geological contact between the Espinhaço Supergroup and the Bambuí Group. As well as considerable biogeographical diversity, the area also encompasses considerable geodiversity, manifested in geological, geomorphological and pedological diversity. Other environmental analysis or mapping studies have been carried out in the contact landscape between the Espinhaço Supergroup and the Bambuí Group (TRAVASSOS; GUIMARÃES; VARELA, 2008; GUIMARÃES; TRAVASSOS; LINKE, 2011; RODRIGUES, 2011; GUIMARÃES, 2012; RODRIGUES; TRAVASSOS, 2013; JANSEN, 2013), but the possible studies have not been exhausted. Therefore, with the aim of continuing research into contact karst, we would like to carry out further analyses based on Tricart (1977); Sotchava (1977); Christofoletti (1999); Monteiro (2001); Crepani et al. (2001); Ab'Saber (2003)[1] ; Bertrand (2004)[2] ; Troppmair and Galina (2006) and Jansen (2013).

Against this backdrop, this work aims to carry out a geoecological analysis of the landscape of a sample area of the geological contact between the Espinhaço Supergroup and the Bambuí Group, considering both the concepts, methodologies and techniques applied and the existing geodynamics as specific objects.

The general objective is to analyse the geoecological conditions of the landscape circumscribed by the boundaries of Topographic Maps SE-23-Z-A-II (Corinto, MG) and SE-23-Z-A- III (Diamantina, MG) through different environmental systems modelling. To this end, the following specific objectives will be developed:

a) Systematise modelling of environmental systems compatible with the geosystemic (SOTCHAVA, 1977; CHRISTOFOLETTI, 1999; MONTEIRO, 2001; BERTRAND, 2004;

---

[1] Collection of articles, in particular chapter 01 - "Mares de Morros", Cerrados e Caatingas: Geomorfologia Comparada, originally published in Mário Guimarães Ferri - *Simpósio sobre o Cerrado*, São Paulo, Edusp, 1963, and chapter 02 - Potencialidades paisagísticas brasileiras, originally published in Recursos Naturais, Meio Ambiente e Poluição, Rio de Janeiro, IBGE/Supren, 1977.

[2] Work originally published in "Revue Geógraphique des Pyrénées et du Sud-Ouest", Toulouse, v. 39, n. 3, p. 249272, 1968, under the title: Paysage et geographie physique globale. Esquisse méthodologique. Translated by Olga Cruz and published in Brazil in Caderno de Ciências da Terra, from the Geography Institute of the University of São Paulo (USP), no. 13, 1972.

TROPPMAIR; GALINA, 2006), morphodynamic (TRICART, 1977), morphoclimatic and phytogeographic (AB'SABER, 2003) approaches, as well as aspects relating to natural and environmental vulnerabilities (CREPANI et al., 2001; JANSEN, 2013).

b) To characterise the main geographical and ecological aspects of the landscape circumscribed by the boundaries of Topographic Maps SE-23-Z-A-II (Corinto, MG) and SE-23-Z-A-III (Diamantina, MG), namely: geological, geomorphological, underground natural cavities, pedological, climatological, hydrographical and phytogeographical (vegetation cover).

c) Carrying out three environmental systems modelling studies to understand the morphodynamics, natural vulnerability to soil loss and natural and environmental vulnerabilities of the landscape circumscribed by the boundaries of Topographic Maps SE-23-Z-A-II (Corinto, MG) and SE-23-Z-A-III (Diamantina, MG), distinguishing between a geosystemic approach and an approach based on morphoclimatic and phytogeographic domains.

d) To validate, with a geosystemic approach, the three models produced through fieldwork in three different sample areas, one located on rocks of the Bambuí Group, another located on the contact between the Bambuí Group and the Espinhaço Supergroup and yet another located on the Espinhaço Supergroup.

1.1 Location and importance of the study area

The area covered by Topographic Maps SE-23-Z-A-II (Corinto, MG) and SE-23-Z-A-III (Diamantina, MG) stretches for 5,834 km^2, covering 55.3 km in a N-S direction and 105.5 km in an E-W direction. According to figure 01, the area covers portions of the territories of the Minas Gerais municipalities of Augusto de Lima, Buenópolis, Corinto, Couto de Magalhães de Minas, Curvelo, Datas, Diamantina, Gouveia, Monjolos, Santo Hipólito and Serro, approximately 250 km north of Belo Horizonte, the state capital.

Access to the area from Belo Horizonte can be gained 1) by following the BR-040 (towards Brasília), BR-135 (towards Corinto) and MG-220 (towards Diamantina) motorways, and 2) by following the MG-010 (towards Serro), BR-259 (towards Datas) and BR-367 (towards Diamantina) motorways.

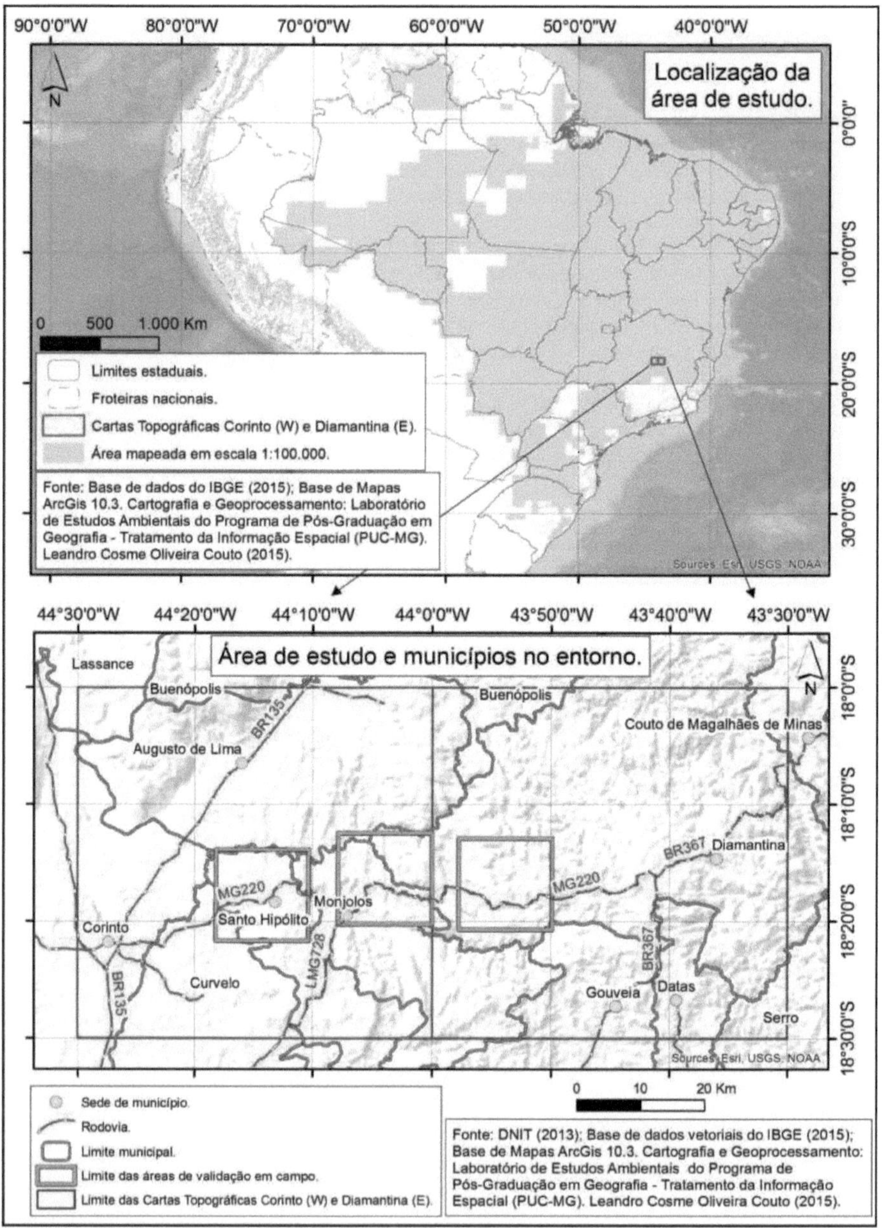

Figure 01 - Maps showing the location of the study area,
Source: Prepared by the author, 10/2015.

The landscape circumscribed by the boundaries of the selected topographic maps is a representative sample of the complexity of various Brazilian geoecological aspects. The following aspects occur mainly in the region:

- The geological contact between the Espinhaço Supergroup and the Bambuí Group, located in the Southern Espinhaço and encompassing considerable geological diversity;

- The contact between a traditional karst region (predominantly calcarenite lithology) and another non-traditional karst region[3] (predominantly quartzite lithology);

- The geomorphological contact between the São Franciscana Depression and Serra do Espinhaço relief compartments, implying a high altimetric gradient;

- The contact between the hydrographic basins of the São Francisco River, through the Rio das Velhas sub-basin, and the Jequitinhonha River;

- The phytogeographic contact between the Cerrado and Atlantic Forest Biomes, implying an anastomosed vegetation cover with grassland, savannah and forest phytophysiognomies.

Becker (2012) states that the facts of space are always unique, since each fact is situated at the intersection of different processes. Therefore, for this dissertation, it is recognised that each of these contacts has geographical and/or ecological relevance on its own, while the combination of their occurrence in the same landscape multiplies the area's relevance for the purposes of environmental study, research, conservation and preservation.

The three areas selected for field modelling validation are georeferenced according to Table 01. Taken together and in sequence, the West, Central and East field validation areas correspond to the realisation of a profile pertinent to the various environmental aspects of the western face of the Serra do Espinhaço.

Table 01 - Boundary coordinates of the areas selected for field modelling validation.

Western Area (over the Bambuí Group)	UTM (SIRGAS2000)	
	Y	X
NW end	7984082,269213	573679,464635
NE end	7984012,885473	587485,631923
SE end	7969627,011123	587406,905054
SW end	7969697,827403	573611,019779
Central Area (on the contact between the Bambuí Group and the Espinhaço Supergroup)	UTM (SIRGAS2000)	
	Y	X
NW end	7986608,513360	591936,129030
NE end	7986539,129620	605742,296318
SE end	7972153,255270	605663,569449
SW end	7972224,071550	591867,684174
Eastern Area (over the Espinhaço Supergroup)	UTM (SIRGAS2000)	
	Y	X
NW end	7985686,374472	609634,509021
NE end	7985616,990732	623440,676309
SE end	7971231,116382	623361,94944
SW end	7971301,932662	609566,064165

---

[3] Terminology used by Andreychuk et al. (2009); Travassos (2010); Guimarães, Travassos and Linke (2011); Jansen (2013).

Figure 02 illustrates the contacts: 1) on the right in the photo (east), in the foreground, rocks of the Espinhaço Supergroup (considered non-traditional karst by some authors), making up a mountainous relief formation with grassland vegetation; 2) in the centre, a traditional karst plain occupied by farming activities and gallery forest; 3) to the left in the photo (west), in the background, rock outcrops of the Bambuí Group (traditional karst), forming a relief formation with escarpments and rock massifs covered by dry forest vegetation.

Figure 02 - NNW view from coordinates 18.28° S and 44.04° W. [4][5]

Studying this landscape, Rodrigues (2011) and Rodrigues and Travassos (2013) state that the region is an extremely important part of Minas Gerais on the international and national stage. Respectively, it contains extensive areas covered by carbonate rocks of the Bambuí Group bordered by the western portion of the Serra do Espinhaço, widely studied in the 19th century by the Danish naturalist Peter Wilhelm Lund (1801-1880), while at the same time it is located at the beginning of the region of the backlands of Minas Gerais portrayed in the literature of the Minas Gerais writer J. Guimarães Rosa (1908-1967).

1.2 Method, techniques and materials

The proposed analyses are carried out using a predominantly deductive method, with a systemic approach, sequenced in the stages shown in Table 02.

Chart 02 - Methodological stages.

STAGES OF THE DEDUCTIVE PROCEDURE	STAGES DEVELOPED IN THIS WORK
Perceptual experiences	Initial fieldwork
Image of real-world structure	
Apriori model	Compilation of approaches and modelling of morphodynamic environmental systems (Tricart, 1977), natural vulnerability to soil loss (CREPANI et al., 2001) and natural and environmental vulnerabilities (JANSEN, 2013), compatible with morphoclimatic and phytogeographic approaches (AB'SABER, 2003) e geosystemic (SOTCHAVA, 1977; TROPPMAIR, 2000a;

---

[4] Conversion of Geographic Coordinates to UTM made using the Geographic Calculator of the National Institute for Space Research (INPE), available at< http://www.dpi.inpe.br/calcula/>.

[5] Enlarged figure included in the Appendix to this dissertation.

	TROPPMAIR, 2000b; MONTEIRO, 2001; BERTRAND, 2004; TROPPMAIR; GALINA, 2006).
Data collection	Consolidation of the digital database
Experimental design (definition, classification, measurement)	Computer modelling of environmental systems and map algebra
Verification procedures	Final fieldwork
Explanation	Geoecological analyses

Source: Adapted from Christofoletti (1999).

The techniques used were fieldwork, computer modelling and map algebra. On 19, 20 and 21 June 2015, fieldwork was carried out in the study area and its surroundings to make initial observations about the location and uniqueness of the landscape. This work focused on the theoretical references of Tricart

(1977), Sotchava (1977), Christofoletti (1979), Monteiro (2001), Bertrand (2004) and Troppmair and Galina (2006), plus Christofoletti (1999) and Ab'Saber (2003).

Following on from the characterisation of the study area's landscape, we sought out other works already carried out, especially those developed within the scope of the Postgraduate Geography Programme at PUC Minas, to continue the research by Travassos, Guimarães and Varela (2008), Guimarães, Travassos and Linke (2011), Rodrigues (2011), Guimarães (2012), Jansen (2013) and Rodrigues and Travassos (2013).

Three environmental modellings were applied using the geoprocessing technique called Map Algebra (or Multicriteria Analysis), applied to the minimum element of raster files (pixel -picture *element)*:

- Modelling morphodynamic environments: adapted from Tricart (1977);

- Modelling natural vulnerability to soil loss: proposed by Crepani et al. (2001) and inspired by Tricart (1977);

- Modelling of natural and environmental vulnerabilities: carried out by Jansen (2013) and adapted from Crepani et al. (2001).

The Map Algebra technique requires the conversion of digital spatial vector data into raster (matrix) format, with the different attributes of the database for each aspect of the landscape converted into their respective mathematical values, which are equated resulting in a final spatial data.

The mathematical values used in the modelling of morphodynamic environments are defined in relative terms based on the theoretical foundation of Tricart (1977): 01 for environmental aspects favourable to morphogenesis, 02 for intermediate environmental aspects and 03 for aspects favourable to pedogenesis. The mathematical values used in the modelling of natural vulnerability to

9

soil loss are those defined by Crepani et al. (2001), who proposed the modelling. Finally, the values and weighting of the different attributes used in the modelling of natural and environmental vulnerabilities follow that proposed by Jansen (2013).

Figure 03 shows a flowchart outlining the environmental aspects used in the map algebra for each of the proposed modellings. The gradual progressive condition of complexity stands out, with the modelling based on Tricart (1977) using three aspects, the modelling based on Crepani et al. (2001), five aspects, and the modelling based on Jansen (2013), six aspects.

Carrying out this research required the use of specialised Geographic Information Systems (GIS) software and consolidated digital data from the study area. The software used was ArcGIS® 10.3, provided by the Environmental Studies Laboratory of the Postgraduate Programme in Geography - Spatial Information Processing at PUC Minas. The digital vector data (shapefile format) used in this work was produced and made available by government organisations via the Internet:

- Planialtimetric data: contour lines produced by the Geographic Service Department (DSG) of the Brazilian Army, for Topographic Chart SE-23-Z-A-II (Corinto, MG), and by the Brazilian Institute of Geography and Statistics (IBGE), for Topographic Chart SE-23-Z-A-III (Diamantina, MG), on a scale of 1:100.000 scale, provided by the State Environment System (Sisema) with geographical projection on the WGS 84 Horizontal Datum; hydrography provided by the National Water Agency (ANA), on a 1:1,000,000 scale with geographical projection on the SAD 69 Horizontal Datum; and roads produced by the National Department of Infrastructure and Transport (DNIT), on a 1:1,000,000 scale, with geographical projection on the WGS 84 Horizontal Datum;

- Geological data (lithology and structure): Companhia de Pesquisa de Recursos Minerais (CPRM), for Topographic Map SE-23-Z-A-II (Corinto, MG), and Companhia de Desenvolvimento de Minas Gerais (CODEMIG) for Topographic Map SE-23-Z-A-III (Diamantina, MG), both in scale 1:100,000 and respectively with geographical projections in Horizontal Datum SAD 69 and WGS 84;

- Speleological data (pinpoint location of natural underground cavities): National Cave Research and Conservation Centre (CECAV) of the Chico Mendes Institute for Biodiversity Conservation (ICMBio), for the national territory, with geographical projection on the SIRGAS 2000 Horizontal Datum;

- Pedological data (soil typology): Department of Soils (DPS) of the Federal University of Viçosa (UFV) for the territory of Minas Gerais, on a scale of 1:650,000 and geographical projection on the WGS 84 Horizontal Datum;

- Phytogeographic data (typology of vegetation cover in 2009): State Institute of Forest (IEF) of Minas Gerais[6] , for the state territory, on a scale of 1:60,000 and geographical projection on the

---

[6] The digital data for mapping the vegetation cover of the state of Minas Gerais in 2009, as well as the contour lines for the state of Minas Gerais, were obtained concurrently from Webgis, available at: http://geosisemanet.meioambiente.mg.gov.br/zee/, Accessed in January 2016.

WGS 84 Horizontal Datum.

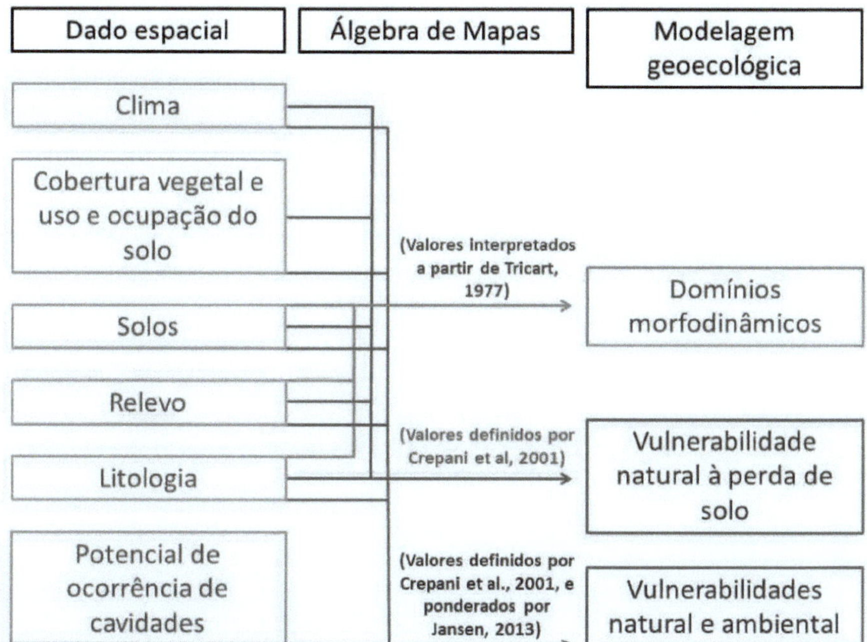

Figure 03 - Map Algebra flowchart.
Source: Prepared by the author.

Mainly with a view to refining phytogeographic and land use and occupation data, images from the Landsat 8 satellite made available by the United States Geological Survey (USGS) for free access via the Internet (http://earthexplorer.usgs.gov/) were also used. This satellite has an operational orbit at a distance of 705 km from the Earth's mean surface, with a temporal resolution of 16 days, a radiometric resolution of 16 Bits, 02 sensors that add up to a total of 11 spectral bands, the OLI sensor for spectral resolution in bands 1 to 8 with 30 metres of spatial resolution, while the TIRS sensor for spectral resolution in band 9 with 15 m and bands 10 and 11 with 100 metres. Each image from this satellite has a width range (coverage area) of 170 by 185 kilometres in the WGS 84 Horizontal Datum geographic projection. In this scenario, images of band 73 were acquired in alignment 218 (Monjolos region, MG), in bands 01 to 9 (highlighting bands 2, blue, 3, green, and 4, red, and 9, panchromatic) dated 08/2014 (dry period, with little cloudiness).

The satellite data was processed to correct topological and database inconsistencies in the vector files, as well as adjusting it to the SIRGAS Horizontal Datum

2000 as established by FIBGE (2005). The satellite image was digitally processed using true colour composition, with the images in bands 2, 3 and 4 channeled into blue, green and red respectively.

Once these procedures had been carried out, the data was considered consolidated, and in parallel the theoretical compilation of environmental systems modelling (Chapter 02) compatible with the morphodynamic approach of Tricart (1977), the *geosystemic* approach of Sotchava (1977) and Bertrand (2004) - cited by Christofoletti (1999), Monteiro (2001) and Troppmair and Galina (2006) - and the *morphoclimatic and phytogeographic* approach of Ab'Saber (2003) was carried out. In addition, aspects of the natural and environmental vulnerabilities of Crepani et al. (2001) and Jansen (2013) were addressed. The consolidated data was then considered suitable for subsequent use in characterising the study area (Chapter 03) and applying the environmental models studied (Chapter 04).

CHAPTER 2

THEORETICAL BACKGROUND

"It's with the 'good' that we get 'bad'. "
(J. Guimarães Rosa.)

2.1.   Geoecological modelling of the landscape.

The effort to carry out geoecological modelling of the landscape requires a clear and integrated understanding of the concepts of model and landscape ecology (or geoecology). Christofoletti (1999) recognises the existence of different nuances in the meaning of these terms and other authors (BARROSO; ABREU, 2003; MONTEIRO, 2001; BERTRAND, 2004) also point out the need to refine the understanding of model and landscape. This dissertation takes as its reference the definition given by Christofoletti (1999, p.08), for whom model:

> (...) in general, it can be understood as 'any simplified representation of reality' or of an aspect of the real world that is of interest to the researcher, which makes it possible to reconstruct reality, predict behaviour, transformation or evolution. (...)

This definition underpins the integrated understanding of the terms "model" and "geoecology" and "landscape", since the aim is to make representations (modelling) of the (geoecological) aspect of the real world (landscape). Christofoletti (1999) bases his definition on the concept of model that is still considered the most appropriate, authored by Haggett and Chorley (1975):

> A model is a simplified structure of reality that supposedly presents important characteristics or reactions in a generalised way. Models are highly subjective approximations because they do not include all the associated observations or measurements, but they are valuable because they obscure accidental details and allow the fundamental aspects of reality to emerge (HAGGETT; CHORLEY, 1975 apud CHRISTOFOLETTI, 1999, p.08).

The highly subjective condition of the models, at the same time "obscuring accidental details" and allowing "fundamental aspects to emerge", justifies the adoption of approaches that theoretically and practically parameterise different landscape modelling, as Bertrand (2004, p.141), a French reference in geosystemic thinking, has done when defining the term landscape:

> (...) the study of landscapes can only be carried out within the framework of a global physical geography.
>
> Landscape is not the simple addition of disparate geographical elements. It is, in a given area of space, the result of the dynamic and therefore unstable combination of physical, biological and anthropic elements which, reacting dialectically to one another, make the landscape a unique and inseparable

whole, in perpetual evolution. (...)

Bertrand's definition underpins the unity of geographical thinking once supported by the thoughts of Paul Vidal de La Blache (1845 to 1918) and Alfred Hettner (1859 to 1941), as Amorim Filho (2006, p. 39) points out:

> (...) both worked with an integrative concept of geographical region, in which the description of the landscape played the main role. This landscape was not just physical, or just human, but rather physical (as an environment that offers itself to the action of society) and human (as works and arrangements produced by society over the course of history). (AMORIM FILHO, 2006, p.39)

In the context of Traditional Geography, landscape is the expression of the area where there is an intimate connection between human beings and the environment, developed over the centuries and forming a singular spatial unit, a region with its own identity. Thus, despite originally dating from 1968, Bertrand's definition expresses the principles indicated by Amorim Filho (2006) as guiding classical or traditional geography, according to the correlation shown in table 03.

According to Christofoletti (1999), the evolution of geographical and ecological knowledge into broader perspectives for characterising nature led to the emergence of numerous proposals for defining and delineating the component units of the earth's surface, most notably the proposal for Landscape Ecology, or Geoecology, made by Carl Troll (1899-1975) in 1938. For this author, both terminologies have the same meaning, in a broad sense, dealing with the simultaneous action and integration of the atmosphere, lithosphere, hydrosphere and biosphere (TROPPMAIR, 2000a). In this proposition, abiotic elements act as conditioning factors for living beings (individuals or groups of organisms / biocenoses) at four different scales of analysis: biome (with greater spatial scope), landscape, ecosystem and, to a lesser extent, population and plants (CHRISTOFOLETTI, 1999).

Chart 03 - Correlation between Bertrand's (2004) definition of landscape and the guiding principles of Traditional Geography indicated by Amorim Filho (2006).

Excerpts from the definition of landscape (BERTRAND, 2004, p. 141)	Guiding principles of classical or traditional geography (AMORIM FILHO, 2006)
"(...) The study of landscapes can only be carried out within the framework of a global physical geography (...)"	1) General Geography and 2) Earth Unit
"(...) And, in a certain portion of space, (...)"	3) Location / position and 4) length
"(...) the result of the dynamic and therefore unstable combination of physical, biological and anthropogenic elements (...)"	5) Connection
"(...) which, reacting dialectically on one another, (...)"	6) Causality
"(...) make the landscape a unique and	7) Differentiation of areas

inseparable whole, (...)"	
"(...) in perpetual evolution (...)"	8) Activity

Source: Prepared by the author.

The geoecological perspective emphasises the need to consider natural landscapes (without anthropogenic action) alongside cultural landscapes (rural and urban) and socioeconomic aspects (CHRISTOFOLETTI, 1999; TROPPMAIR, 2000b), and is applied to regional, rural and urban planning. In this scenario, the morphoclimatic and phytogeographic (AB'SABER, 2003), geosystemic (SOTCHAVA, 1977; MONTEIRO, 2001; BERTRAND, 2004) and morphodynamic (TRICART, 1977; CREPANI et al, 2001; JANSEN, 2013) approaches are relevant.

2.2.   Morphoclimatic and phytogeographic approach.

In line with the geoecological perspective summarised by Christofoletti (1999) and Troppmair (2000b), AB'SABER (2003) notes that landscape is:

-   Both a legacy of ancient physiographic and biological processes (millions to tens of millions of years), remodelled and modified by recent processes (a few thousand to tens or even hundreds of thousands of years);

-   As the collective heritage of the peoples who have historically inherited them as the territory in which their communities operate.

The general compartmentalisation of the topography has not been significantly modified by recent processes (with regional or local effects), which has resulted in the dominance of a

global scheme of landscapes, zonal and azonal, very close to the landscape picture of the earth's surface still seen today by the peoples who have historically inherited them as territory (AB'SABER, 2003). Thus, when studying the broad and diverse Brazilian landscape, this author establishes the concept of morphoclimatic and phytogeographic domain as:

> (...) a spatial ensemble of a certain order of territorial magnitude - from hundreds of thousands to millions of square kilometres - where there is a coherent scheme of relief features, soil types, vegetation forms and climatic-hydrological conditions. These spatial domains of integrated landscape and ecological features occur in a kind of main area, of a certain size and arrangement, in which the physiographic and biogeographic conditions form a relatively homogeneous and extensive complex. To this more typical and continuous area - as a rule, with a polygonal arrangement - we apply the name of core area, soon translated as *nuclear* area (...) (AB'SABER, 2003, pp. 11-12)

Brazil has six major landscape domains (four intertropical and two subtropical) based on morphoclimatic and phytogeographic conditions, as shown in Table 04:

15

Table 04 - Brazilian landscape domains according to Ab'Saber (2003).

Landscape	Morphoclimatic and phytogeographic conditions	
Amazon	Equatorial forested lowlands	Intertropical
Cerrado	Tropical inland plateaus with savannahs and forests-gallery	
Mares de Morros	Forested tropical-Atlantic mammillary areas	
Caatingas	Semi-arid intermontane and interplateau depressions	
Araucarias	Subtropical plateaus with Araucaria trees	Subtropical
Prairies	Subtropical hillsides with mixed grasslands	

Source: Prepared by the author.

According to AB'SABER (2003), these core areas, in polygonal arrangements, are typically zonal, with the exception of the Mares de Morros, and interspersed with elongated transition and contact strips, which are clearer in terms of the environmental components of vegetation and soil and less clear in terms of relief. Figure 04 shows an illustration of the map of Brazilian morphoclimatic domains, originally drawn up by AB'SABER:

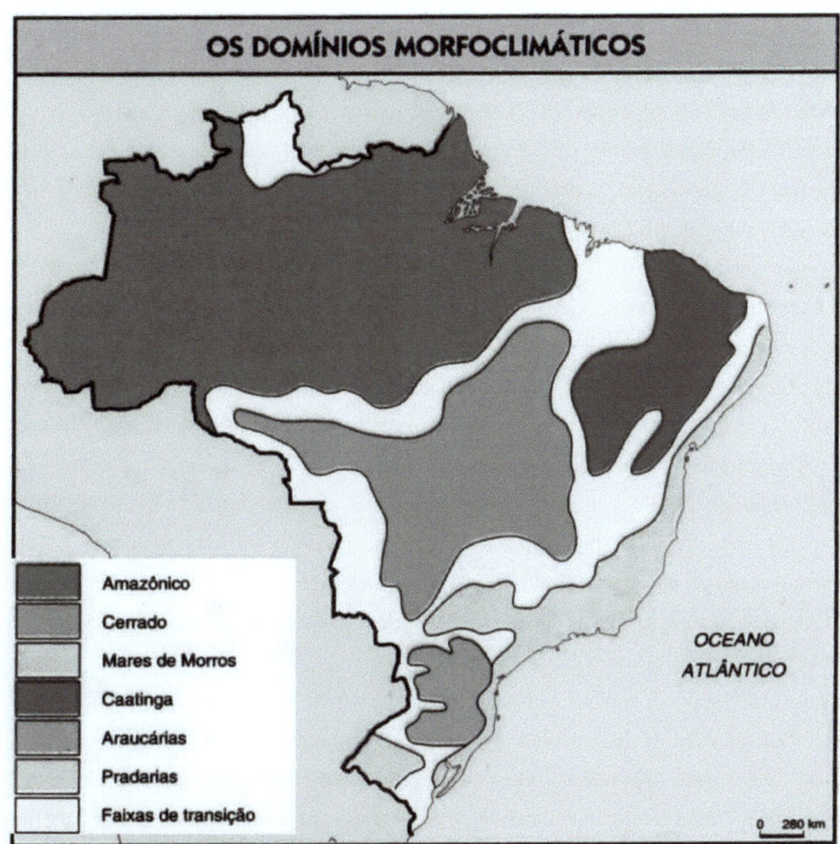

Figure 04 - Map of Brazilian morphoclimatic domains.
Source: DOMINIOS..., 2016.

These "transitional interspaces" present direct combinations between the environmental components of the nuclear areas that surround them, resulting in anastomosed landscapes with singular sectors, sometimes azonal. It is possible for azonal enclaves of one type of domain to occur on the edges of the neighbouring nuclear area, as well as relictual enclaves within the neighbouring nuclear area, explained by local factors of morphoclimatic or geobotanical exception.

The area under study in this dissertation is part of the transition between the azonal domains of the forested Mares de Morros and the zonal savannah plateaus of the Cerrado penetrated by gallery forests. Its study allows for a better scientific understanding of the anastomosed (contact) landscape between these domains, but requires a scale of landscape analysis with less spatial coverage than the morphoclimatic and phytogeographic domains. In this scenario, the geosystemic approach appears pertinent to the geological modelling intended for the Monjolos and Diamantina landscapes.

2.3.   Geosystem approach

Christofoletti (1999), Monteiro (2001), Troppmair and Galina (2006) and Pissinati and Archela (2009) recognise the use of the term "geosystem" as having a history that began Ann in the 1960s and in which the Soviet V. B. Sotchava (1977), representative of the Siberian school, and the French P. G. Bertrand (1935), representative of the French Pyrenees school, stand out. Monteiro (2001) highlights the fact that geosystemic concerns converge in the integrated nature studies of these two schools of thought, although this approach arose independently of contact between the two.

The term "geosystem" was introduced by Sotchava in 1962 (CHRISTOFOLETTI, 1999; MONTEIRO, 2001; TROPPMAIR; GALINA, 2006), designating the Geographical System or Territorial Natural Complex. For Sotchava (1977), a geosystem is a peculiar class of open and hierarchically organised dynamic systems. Reflecting the Soviet context of vast territorial extensions (TROPPMAIR; GALINA, 2006), Sotchava (1977) established the geosystem as a natural phenomenon in which economic and social factors would affect it, creating spatial peculiarities.

From Sotchava's (1977) perspective, the geosystem can be classified bilaterally into elementary spatial units: geomers (homogeneous areas) and geocores (differentiated areas). The elementary geomers and geocores are delimited by the horizontal extent to which the natural elements ensure the same systemic unit, as well as the maximum vertical extent of 50 metres in the atmosphere (SOTCHAVA, 1977).

According to Troppmair and Galina (2006), Sotchava left the term "geosystem" rather vague and flexible. This allowed it to be used by different geographers with different contents, methodologies, scales and approaches. Most notably, P.G. Bertrand incorporated the term "geosystem" in 1968, after a few years of studying the integration of environmental elements in the landscape.

According to Bertrand (2004), a taxonomy of the landscape should not be made by imposing pre-established categories, but by researching the objective discontinuities of the landscape. From Bertrand's perspective, landscape taxonomy has six different temporal-spatial levels depending on the scale, with the climatic and structural relief elements underpinning the recognition of the higher units (of greater spatial scope) Zone, Domain and Natural Region, and the biogeographical and anthropogenic elements underpinning the recognition of the lower units (of lesser spatial scope) Geosystem, Geofacies and Geotopes. In this scenario:

> The geosystem is located between the 4th and 5th temporal and spatial magnitudes[7]. It is therefore a dimensional unit between a few square kilometres and a few hundred square kilometres. It is on this scale that most of the phenomena of interference between the elements of the landscape take place and where the dialectical combinations that are most interesting to the geographer evolve. At the higher levels, only the relief and climate matter and, incidentally, the large plant masses. At lower levels, biogeographical elements are capable of masking the overall combinations. In short, the geosystem is a good basis for studying the organisation of space because it is compatible with the human scale.

---

7 Cailleux, A. and Tricart, J. **Le problème de la classification des faits géomorphologiques**. Ann. de Géogr., 65:162 -186. 1956.

(BERTRAND, 2004, p. 146)

In an outline of a theoretical definition, Bertrand (2004) defines the Geosystem as the result in the landscape of the combination of: 1) ecological potential (nature of the rocks and surface mantles, dynamics and value of the slope of the slopes, dynamics of precipitation, temperature, water tables, springs, value of the pH levels of the water and drying times of the soil, etc.); 2) biological exploitation (living communities of plants and animals in ecosystem dynamics); and 3) anthropic action (socio-economic activities).

Although Bertrand (2004) recognises that within a geosystem there is relative ecological continuity and that every transition to another geosystem is characterised by an ecological discontinuity, he also recognises that the geosystem does not necessarily present itself with great physiognomic homogeneity. Due to internal dynamics (various stages of evolution of its component elements), different landscapes predominantly form. Thus, "(...) geo 'system' emphasises the geographical complex and the dynamics of the whole; geo 'facies' insists on the physiognomic aspect and geo 'top' places this unit at the last level of the spatial scale" (BERTRAND, 2004, p.145).

For Christofoletti (1999), Sotchava and Bertrand's propositions are dedicated to establishing a specific scale of magnitude for the geosystem through a spatial hierarchy, from the local to the global terrestrial scale. Just as Sotchava and Bertrand's geosystemic perspectives differ in that they respectively reflect the authors' experiences in Soviet and European territories (MONTEIRO, 2001; TROPPMAIR; GALINA, 2006), the works reported by Troppmair (2000b), Monteiro (2001) and Pissinati and Archela (2009) present geosystemic perspectives that reflect the authors' experiences in Brazilian territory, with greater or lesser proximity to the propositions of Sotchava and, predominantly, Bertrand.

This dissertation uses the morphoclimatic / phytogeographic (AB'SABER, 2003) and geosystemic (BERTRAND, 2004) approaches to characterise the study area. In all these perspectives, terrestrial morphology acts as a reference for classifying the landscape, which is why Tricart's (1977) morphodynamic approach is also pertinent.

2.4.   Morphodynamic Approach

The morphodynamic approach deals with the role of the morphogenic component in the dynamics of the earth's surface and was formally introduced in Brazil by the Frenchman Jean Tricart through the study of landscape ecodynamics.

According to a presentation by a representative of the IBGE[8] (TRICART, 1977), Tricart's study emphasises the discipline of good use of natural goods and wealth for the benefit of the community, using the systems approach as something necessary for understanding the environment. For Tricart (1977, p. 32), "the use of the logical instrument of systems makes it possible to quickly identify the indirect changes that will be triggered by an intervention that affects this or that other element of the ecosystem", allowing society to make decisions based on land-use planning and management criteria.

Tricart (1977) considers the morphogenic component to be the most important in the dynamics of the earth's surface, since morphogenic processes produce surface instability and this is a limiting factor for the development of living beings. Thus, Tricart (1977) emphasises the need to establish a taxonomy of environmental types based on the degree of morphodynamic stability-instability. He therefore proposes a methodology based on the study of the dynamics of ecotopes (the environment of an ecosystem), which he calls ecodynamics, and which distinguishes three major types of morphodynamic environments according to the intensity of the natural processes involved in morphogenesis and pedogenesis.

Tricart's proposition (1977) distinguishes between stable and strongly unstable environments in the landscape, with opposing behaviours, and integrated environments, with behaviour intermediate to the other two. Each of the three types of environments encompasses various areas of the earth's surface that are similar in terms of morphodynamic behaviour (which is why the term is used in the plural):

a) **Stable environments** have a slow evolution of the landform, denoting geomorphological stability, either over a long period of time (rarer) or relatively recently, beginning in the Holocene (recurrent). The effects of pedogenesis outweigh the effects of morphogenesis, with polyphase soils and little dissection of the relief. There is little influence from lithological conditions, which together with bioclimatic conditions contributes to the predominance of aged (ancient) soils and landforms. For environmental conservation in these environments, it is necessary to maintain vegetation cover of a density equivalent to climax vegetation; for agropastoral use, it is necessary to specifically study the soil in situ, which varies in space, in order to act with improvements and fertilisations.

b) **Strongly unstable environments** have a considerable predominance of morphogenesis over pedogenesis, with rugged relief (dissected forms with steep slopes and carved valleys) and recent and lithic soils, both of which are decisively influenced by lithological conditions. Localised morphogenetic phenomena occur, either sporadically (mass movements of mudflows or falling blocks) or recurrently (ravines and floodplains), and are widespread (diffuse surface runoff, floodplains). Environmental conservation is necessary to protect land and watercourses downstream; agricultural and pastoral uses are marginal and should be avoided because of the risk of irreversible environmental degradation.

---

[8] Geographer Miguel Alves de Lima.

c) ***Intergraded* environments** encompass the gradual transition between stable and highly unstable **environments**. In these environments, there is permanent competition between pedogenesis and morphogenesis, which exert reciprocal interference, with one overpowering the other in an insensitive way and according to specific local situations. When pedogenesis predominates, morphogenesis interferes in the evolution of soils, rejuvenating them to the point of preventing them from becoming old and tending towards geomorphological stability; when morphogenesis predominates, soils are affected by diffuse surface runoff and/or mass movements, tending towards geomorphological instability. Both soils and landforms are of relatively intermediate age and make up mosaics in these landscapes. Longer-term climatic fluctuations (e.g. drier or wetter periods) or anthropogenic interventions (e.g. agriculture and livestock farming) can affect the balance of this reciprocity, culminating in the maintenance of stability or the induction of geomorphological instability. Thus, for environmental conservation it is necessary to maintain vegetation cover in order to inhibit landslides and/or ravines; for agropastoral use it is appropriate to maintain at least herbaceous vegetation cover concomitant with hydraulic management of runoff, avoiding erosion, as well as preventing soil compaction due to cattle trampling.

Since "studying the organisation of space means determining how an action fits into the natural dynamic, in order to correct certain unfavourable aspects and to facilitate the exploitation of the ecological resources that the environment offers" (TRICART, 1977, p. 35), the taxonomy proposed by the author allows for the geoecological modelling of the landscape in three different morphodynamic domains. To this end, considering in particular the structure and content of the synthesis table presented by Tricart (1977) at the end of his work Ecodynamics, the following assumptions are made:

a) The performance of morphogenesis is measured by the relief slope variable, which is parameterised by grouping the 06 classes presented by IBGE (2007) into 03: 0 to 08% (flat and gently undulating classes), 08.01 to 45% (undulating and strongly undulating classes) and greater than 45% (mountainous and steep classes);

b) The performance of pedogenesis can be gauged by the soil maturity variable, parameterised by grouping the 14 soil types presented by the IBGE (2007) into 03 groups, according to the adaptation of the proposal by Crepani el al. (2001): recent (rocky outcrops, gleissolos, neossolos, organossolos, plintossolos and vertissolos), rejuvenated (argissolos, cambissolos, chernossolos, espodossolos, luvissolos, nitossolos and planossolos) and aged (latossolos);

c) The influence of lithology is measured by the rock cohesion variable, parameterised by grouping the most common rock vulnerability to denudation scale (CREPANI el al., 2001; JANSEN, 2013) into three groups: low (01 to 1.6), medium (1.7 to 2.3) and high (2.4 to 3.0).

"(...) The degree of cohesion of rocks is understood to be the intensity of the bond between the rocks.

minerals or particles that make them up" (CREPANI et al., 2001, p.13).

> (...) "The degree of cohesion of rocks is the basic geological information to be integrated from Ecodynamics, since erosive processes that modify landforms (morphogenesis) can prevail in rocks that are not very cohesive, while weathering processes and soil formation (pedogenesis) should prevail in rocks that are very cohesive." (CREPANI et al., 2001, p.73)

Each of these three variables is valued between 01 for the most stable parameter, 02 for the intermediate and 03 for the most unstable, as shown in table 05.

Table 05 - Parameters and values for morphodynamic modelling of the landscape based on Tricart (1977).

Variable	Parameters	Value	Parameters	Value	Parameters	Value
Relief slope (%)	0a8	01	8 a 45	02	>45	03
Soil maturity	Aged	01	Rejuvenated	02	Recent	03
Vulnerability to rock denudation	Reduced	01	Median	02	High	03
Morphodynamic domain	Stable	03	*Inerrgrade*	06	Strongly Unstable	09

Source: Prepared by the author.

Digital modelling takes place by calculating the sum of the values of the three variables in each pixel[9] , resulting in the values 03, 06 and 09 respectively for the morphodynamic domains of stable, intergrade and strongly unstable.

$$MM = IR + MS + GCR$$

Where:

MM = Morphodynamic medium;
IR = Relief slope;
MS = Soil maturity;
GCR = Degree of rock cohesion.

The term *intergrade* means transition: "These media, in effect, ensure the gradual passage between stable media and unstable media. The rubric is, as a matter of fact, conventional because there is no cut; on the contrary, we are in the presence of a continuum" (TRICART, 1977, p.47). Thus, corresponding to this transitional continuum, as well as the variations that exist within the same morphodynamic environment, the eventual results with values 05 and 07, respectively, denote more stable or more unstable intergraded areas, while the results with values 04 and 08 denote stable and unstable areas, according to variation A in figure 05.

---

[9] Run in ArcGis 10.2 *software* using the following tool: *ArcToolbox > Spatial Analyst Tools > Map Algebra> Raster Calculator.*

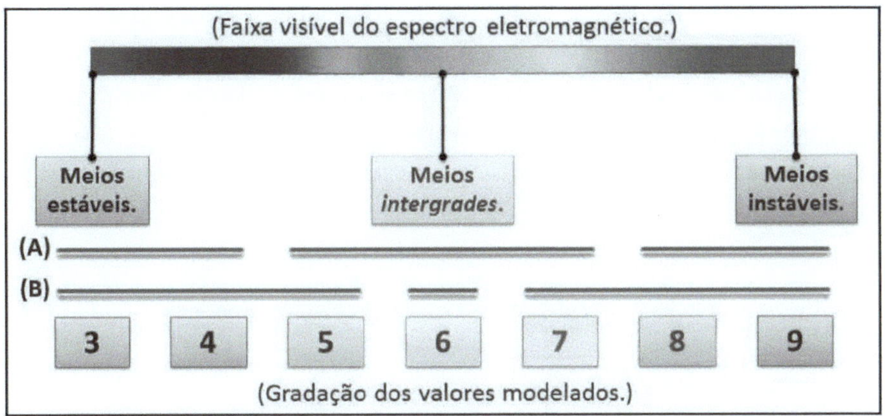

Figure 05 - Gradation of the modelled values in the inorphodynamic continuum.

Source: Prepared by the author.

Expressed in variation B of figure 05, another possible interpretation of the gradation of values modelled in the morphodynamic continuum recognises the intergrade environment as a threshold between stability and instability, which vary in intensity between moderate and strong, as shown in table 06:

Chart 06 - Intensity variations **in the inorphodynamic continuum.**

Modelled values	Morphodynamic environments	Intensity variations in the continuum
03		Strong
04	Stable	Median
05		Moderate
06	*Intergrade*	Weakly stable, on the brink of stability
07		Moderate
08	Unstable	Median
09		Strong

Source: Prepared by the author.

Considering that Tricart (1977) indicates general consequences for environmental conservation and agronomic use in each type of morphodynamic environment, it is feasible to compare current land use and occupation with these indications in order to verify the geoecological perspectives of the landscape.

Other environmental modelling derived from Tricart's methodology (1977) was consolidated by Crepani et al. (2001) and Jansen (2013) and focused on natural vulnerability to soil loss and natural environmental vulnerability, capable of contributing to a systemic understanding of the organisation of space on the earth's surface.

2.5. Natural Vulnerability to Soil Loss

Based on the concept of Ecodynamics and the use of satellite images, which allowed the reinterpretation of pre-existing thematic data in a synoptic and holistic approach, the work by Crepani et al. (2001) presents a methodology for identifying and mapping the natural vulnerability to soil loss in the landscape of the Legal Amazon as a subsidy for the preparation of Ecological-Economic Zoning.

In the model proposed by Crepani et al. (2001), the landscape is classified into two types of overlapping basic territorial units (BTU): one or more "natural landscape units", which represent the natural physical and biotic environmental conditions, overlapped by an "anthropogenic intervention polygon", which represents human action on environmental conditions. In the authors' terms:

> Knowledge of the mechanisms at work in natural landscape units makes it possible to guide the activities to be carried out within the polygon of anthropic intervention, in order to avoid irreversible damage and achieve greater productivity, as well as directing corrective actions within those polygons where inappropriate use causes disastrous consequences. (CREPANI et al., 2001, p. 15)

The principles of Tricart's Ecodynamics (1977) are manifested in the model by Crepani et al. (2001):

a) In the understanding that "(...) when morphogenesis predominates, erosive processes prevail, modifying landforms, and when pedogenesis predominates, soil-forming processes prevail" (CREPANI et al., 2001, p.13);

b) The distinction of 21 different natural landscape units in the morphodynamic continuum, which are distributed between situations of predominance of pedogenesis, at the extreme of stability, and morphogenesis, at the extreme of instability, passing through intermediate situations.

With the pixel as the elementary spatial unit, digital modelling of natural vulnerability to soil loss was established using the arithmetic mean of values corresponding to the influence of physical and biotic environmental aspects (geology, geomorphology, pedology, vegetation and climate) on the relationship between morphogenesis and pedogenesis.

Scaled by Crepani et al. (2001) with values close to 01 denoting the prevalence of pedogenesis and values close to 03 dealing with the prevalence of morphogenesis, the environmental aspects and their respective attributes are:

- *Lithology:* degree of rock cohesion (GCR);
- *Relief:* dissection of the relief by drainage (DD), altitude range (AA) and slope (D);
- *Soils:* development or maturity of the soil (MS);
- *Vegetation:* density of vegetation cover (DCV);
- *Climate:* erosivity (E), resulting from rainfall.

Thus, the natural vulnerability to soil loss (VNPS) is calculated[10] by Crepani et al. (2001) with

---

[10] Run in ArcGis 10.2 *software* using the following tool: *ArcToolbox* > *Spatial Analyst Tools* > *Map Algebra> Raster Calculator.*

the equation :[11]

$$VNPS = (GCR + DD + AA + D + MS + DCV + E) / 7$$

Crepani et al. (2001) state that rocks with a low degree of cohesion between minerals favour the occurrence of morphogenetic processes, just as highly cohesive rocks favour pedogenesis processes. Thus, low soil maturity indicates the prevalence of erosive processes of morphogenesis where young soils occur; in turn, soil maturity (leached and well-developed) implies conditions of stability in which soil formation predominates.

Morphogenetic processes prevail in natural landscape units with a high amplitude of relief, slope and degree of dissection, since there is greater potential energy (CREPANI et al., 2001). Where these morphometric characteristics are reduced, pedogenetic processes prevail.

Vegetation cover provides morphodynamic protection for natural landscape units, so that morphogenetic processes are predominant in lower density vegetation cover (land cover), while "pedogenetic processes occur in situations where denser vegetation cover allows the soil to develop and mature" (CREPANI et al., 2001, p.15). In contrast to vegetation cover, higher erosivity, due to higher rainfall intensity, favours morphogenesis, while lower erosivity, due to lower rainfall intensity, favours pedogenesis.

The scaled values oscillate between 01 and 03, being relative and empirical, assigned separately by the authors to each environmental aspect, and making up a scale of 21 levels of vulnerability of landscape units to soil loss, as shown in Table 08:

---

[11] For the sake of better specification, the acronyms used here refer to attributes, as opposed to those used by Crepani *et al.* (2001), which generally refer to environmental aspects.

Unidades de Paisagem	Média			Grau de Vulnerabilidade	Grau de saturação			
					Verm.	Verde	Azul	Cores
U1		3,0		Vulnerável	255	0	0	
U2		2,9			255	51	0	
U3		2,8			255	102	0	
U4	V	2,7			255	153	0	
U5	U	2,6		Moderadamente Vulnerável	255	204	0	
U6	L	2,5	E		255	255	0	
U7	N	2,4	S		204	255	0	
U8	E	2,3	T		153	255	0	
U9	R	2,2	A	Medianamente Estável/ Vulnerável	102	255	0	
U10	A	2,1	B		51	255	0	
U11	B	2,0	I		0	255	0	
U12	I	1,9	L		0	255	51	
U13	L	1,8	I		0	255	102	
U14	I	1,7	D	Moderadamente Estável	0	255	153	
U15	D	1,6	A		0	255	204	
U16	A	1,5	D		0	255	255	
U17	D	1,4	E		0	204	255	
U18	E	1,3		Estável	0	153	255	
U19		1,2			0	102	255	
U20		1,1			0	51	255	
U21		1,0			0	0	255	

Table 08: Scale of vulnerability of landscape units.
Source: CREPANI et al. (2001, p.21).

2.6. Natural and Environmental Vulnerabilities

Based on the model by Crepani et al. (2001), Jansen (2013) developed his own model focused on the environmental vulnerability of the Morro da Pedreira Environmental Protection Area and the Serra do Cipó National Park for the protection of its speleological heritage. The modelling proposed by Jansen (2013) encompasses two stages very similar to Bertrand's (2004) geosystem approach, with the first resulting from the weighted map algebra of 05 environmental variables typical of ecological potential. The second is the result of simple map algebra between the result of the first stage and the vegetation cover variable, which is typical of the biological exploitation of the landscape and which, together with land use and occupation, also refers to anthropogenic action.

In addition to using the attributes topographic roughness and rainfall index to replace the morphometric attributes and erosivity defined by Crepani el al. (2001), Jansen (2013) initially removes the environmental aspect vegetation and adds the environmental aspect speleological potential to calculate the natural vulnerability of the landscape using the sum weighted :[12]

$$VN = GEO*0.30 + PED*0.25 + ICR*0.20 + IP*0.15 + PCAV*0.10$$

Where:

VN = Natural Vulnerability;
GEO = Geology;
PED = Pedology;
ICR = Topographic Roughness Concentration Index;
IP = Rainfall Intensity;
PCAV= Potential Occurrence of Cavities.

The potential for the occurrence of natural underground cavities was parameterised by Jansen (2013) into the following classes: very high, high, medium, low and unlikely occurrence, which correspond to the methodological terms of the Map of Potential Occurrence of Caves in Brazil, on a scale of 1:2,500,000 produced by CECAV (2012) .[13]

Although based on the methodologies of Meneses (2003) and Gomes (2010), both related to speleological heritage, the weighting was arbitrary and refined on the empirical basis of the author and the National Centre for Cave Research and Conservation (CECAV). The resulting VN varies between 09 and 36 and is categorised into five relative classes established by the Natural Breakdown statistical method, which are: very high, high, medium, low and very low. Jansen (2013) then takes up the environmental aspect of vegetation, adding it to the NPV to consolidate environmental vulnerability:

$$VA = (VN*0.50) + (VEG*0.50)$$

Where:

VA = Environmental Vulnerability;
VEG = Vegetation.

The resulting VA is also categorised into 5 relative classes (very high, high, medium, low and very low) established by the Natural Breakdown method and subsidises the protection of speleological heritage in areas susceptible to anthropogenic actions. In this way, the results of Jansen's model (2013) differ from the results of modelling based directly on Tricart (1977) and subsidise environmental conservation and agricultural uses. The model by Crepani et al. (2001) favours

---

[12] Executed in ArcGis 10.2 software using the tool: ArcToolbox > Spatial Analyst Tools > Overlay > Weighted Overlay .

[13] Available at: http://www.icmbio.gov.br/cecav/projetos-e-atividades/potencialidade-de-ocorrencia-de-cavernas.html.

ecological-economic zoning and land management. All of these models involve a geosystemic approach and evoke the various geographical principles mentioned above. However, each one makes a different contribution to the geoecological understanding of the landscape.

Christofoletti (1999) and Monteiro (2001) point out the misconceptions of Brazilian environmental studies, which section off and dissociate environmental aspects sequentially without thematic integration. In this scenario of the necessary integration of different environmental aspects, the approaches contained in the Theoretical Foundation, combined with the concern to comply with the guiding principles of Traditional Geography, guide the environmental characterisation of the study area.

CHAPTER 3

CHARACTERISATION OF THE STUDY AREA

> *"It is the past that has come to me, like a cloud, to be recognised, only I don't know how to decipher it. " (J. Guimarães Rosa.)*

The study area, delimited by the borders of the topographic maps of Corinto to the west and Diamantina to the east, was visited during the initial fieldwork, carried out on 19, 20 and 21/06/2015, and the final fieldwork, carried out on 18, 19 and 20/02/2017. The initial fieldwork took place along the MG 220 motorway, from west to east, starting from the municipal seat of Corinto, passing through the municipalities of Santo Hipólito and Monjolos and entering Diamantina up to its municipal seat (Figure 06).

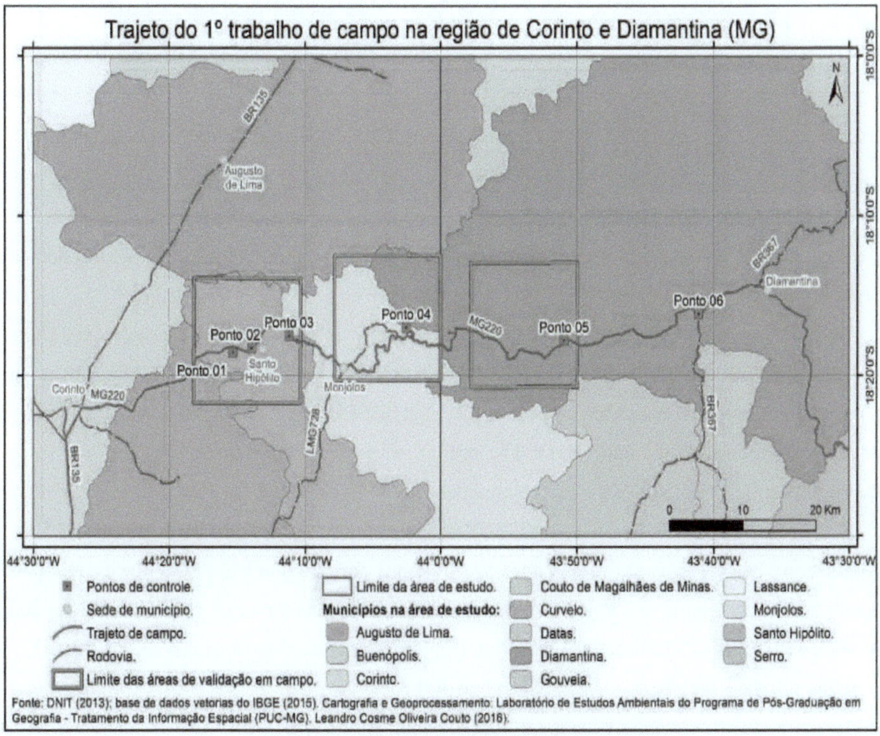

Figure 06 - Map of the initial fieldwork route.
Source: Prepared by the author.

In the initial fieldwork, 6 control points were recorded, distributed among the 3 areas (west, central and east) where the geoecological modelling to be applied to the region's landscape was validated in the field. The points are located along or near the MG

220 in distinctly different conditions in terms of geological, geomorphological, pedological and vegetation cover, as shown in Table 07:

Chart 07 - Control points from the initial fieldwork.

Control Point	UTM coordinates (SIRGAS2000)		Geoecological modelling validation area	General description
	X	Y		
01	578.600	7.975.489	West	Former soil borrowing area near the Rio das Velhas
02	581.055	7.975.980	West	MG 220 - Road bridge over the Rio das Velhas
03	586.019	7.977.344	West	MG 220 - Neossol profile under dry forest
04	601.239	7.978.162	Central	Lithological contact between the Bambuí Group and the Espinhaço Supergroup
05	621.806	7.976.580	East	Flattening between the tops of the Serra do Espinhaço
06	639.099	7.979.526	Complement to the East area	Road junction between MG 220 and BR 367

Source: Prepared by the author.

Control Point 06 has similar environmental conditions to Control Point 05, but is located in a different catchment area, which allows it to be considered as a complement to the validation of Control Point 06.

The study area covers stretches of the Water Resources Planning and Management Units (UPGRH) Rio das Velhas (sub-basin of the São Francisco River) and Alto Jequitinhonha (upstream portion of the Jequitinhonha River basin, part of the so-called Eastern Atlantic Basins), respectively designated by the codes (SF5) and (JQ1). The region is crossed longitudinally by the interfluve between these two UPGRH, with most of it corresponding to stretches of the Lower Rio das Velhas drained by the Rio das Velhas itself (in a SE-NW direction) and by the right bank tributaries Rios Pardo Grande and Pardo Pequeno (both in an E-W direction), forming the Rio Pardo; the eastern part of the study area corresponds to a stretch of the Upper Jequitinhonha River (S-N direction), drained by the Pinheiro River (SW-NE direction) and other tributaries on the left bank of this river (Figure07).

Although the Ribeirão das Varas is just one of the tributaries of the Rio Pardo Grande, it was included as the main watercourse because it was relevant to this study as it runs through the field validation areas. In addition to this and other initial observations about the location and importance of the landscape, and using the morphodynamic and geosystemic approaches as theoretical references, plus the morphoclimatic and phytogeographic approach, the environmental differences between the travelled portions of the São Francisco and Jequitinhonha river basins were verified.

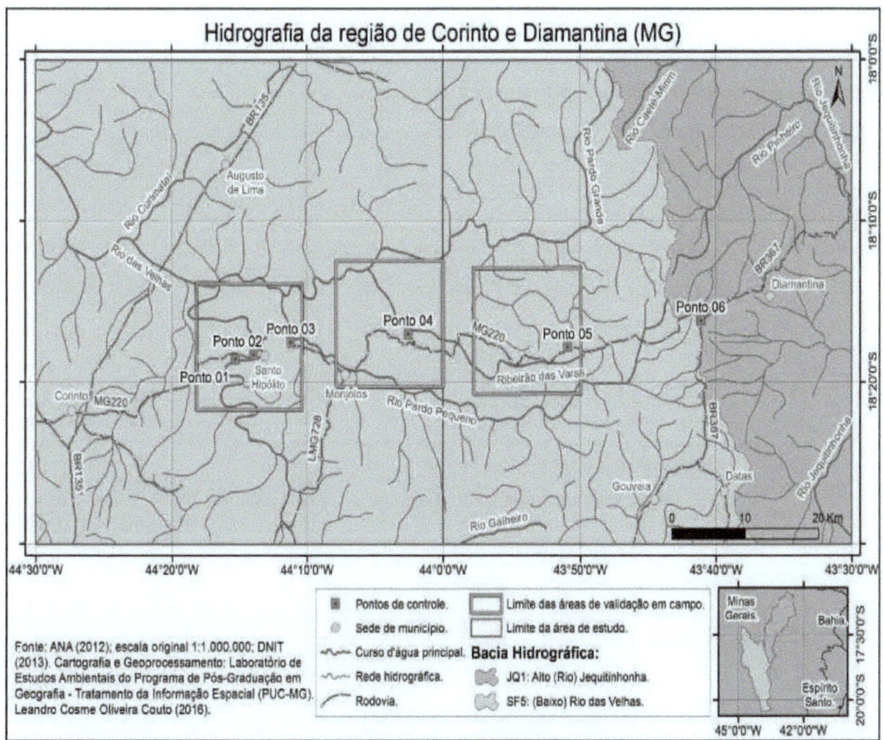

Figure 07 - Hydrographic map of the study area.
Source: Prepared by the author.

Analysing the six control points along the route, the following variations were recorded:

- Lithological distribution in the geological contact between the Espinhaço Supergroup and the Bambuí Group;

- Geomorphological features between the São Franciscana Depression and Serra do Espinhaço relief compartments;

- Distribution of soil and vegetation cover in the phytogeographic contact between the Cerrado and Atlantic Forest biomes.

Subsequently, these records were correlated with existing data, studies and mapping of the region (TRAVASSOS; GUIMARÃES; VARELA, 2008;

GUIMARÃES; TRAVASSOS; LINKE, 2011; RODRIGUES, 2011; GUIMARÃES, 2012; RODRIGUES; TRAVASSOS, 2013; JANSEN, 2013).

3.1. Geological aspects

The study area encompasses lithological diversity in which, to the west, the São Francisco Supergroup is represented by the Bambuí Group, subdivided into the Serra de Santa Helena Formation (argillites and siltstones) and the Lagoa do Jacaré Formation (limestones and marls), and to the east are the rocks of the Espinhaço Supergroup (quartzites, phyllites, metargillites and metasiltites).

These are the São Francisco Craton, covered by rocks of the Bambuí Group, and the Araçuaí Fold Belt, represented by the folds of the Serra do Espinhaço Orogenic Belt (ABREU, 1995; SAADI, 1995; UHLEIN; TROMPETTE; EGYDIO-SILVA, 1995; KNAUER et at., 2011; RODRIGUES, 2011; GUIMARÃES, 2012; FOGAÇA, 2012).

> The structural unit of the São Francisco Craton is made up, stratigraphically, of the unremobilised Archaean/Palaeoproterozoic-age basement in the Belo Horizonte region; the Espinhaço Supergroup in the Serra do Cabral region; and the Jequitaí Formation and the Bambuí Group, the latter two generally included in the São Francisco Supergroup. (...)
>
> The Araçuaí Fold Belt can be subdivided into two structural domains (...) The external unit of the Araçuaí Belt is made up of the basement remobilised by the Brasilian orogeny (Gouveia, Porteirinha and Guanhães Anticlinories), the Espinhaço Supergroup of the homonymous mountain range (southern sector) and the Macaúbas Group (including the Salinas Complex or Unit, according to Pedrosa Soares et al. 1992).
>
> The internal unit of the Araçuaí Belt is the northern extension of the Atlantic Belt or Ribeira Belt (...) (UHLEIN; TROMPETTE; EGYDIO-SILVA, 1995, p. 99).

The study area is predominantly characterised by metasedimentary rocks (quartzites) along the entire length of the Diamantina Topographic Map (E) and sedimentary rocks (argillites, siltstones, conglomerates, limestones and marls) on the Corinto Topographic Map (W), as shown in figure 08. On the eastern edge and the north-western portion of the Corinto Chart, quartzites occur in a longitudinal band (respectively the extension of the quartzites on the Diamantina Chart and the southern part of the Serra do Cabral).

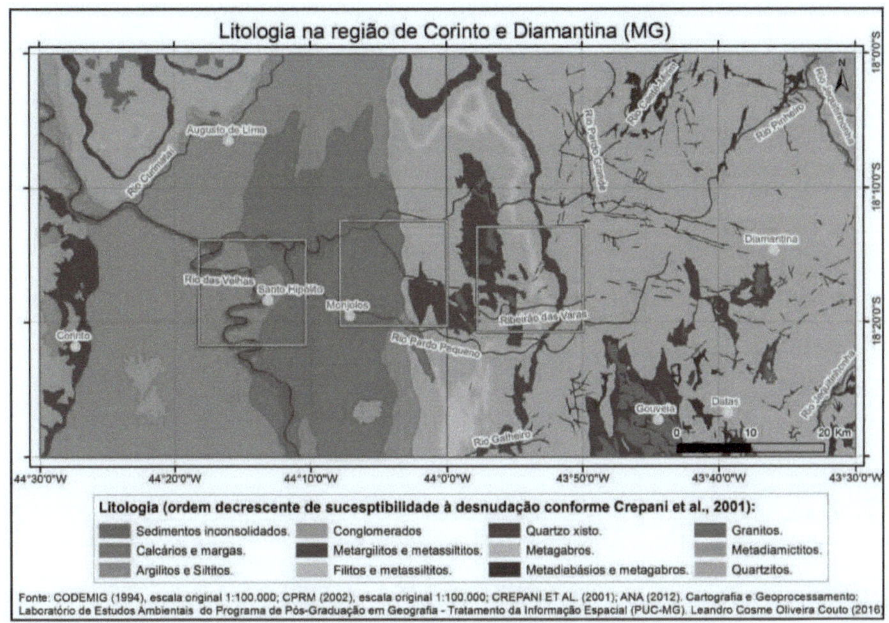

Figure 08 - Lithological map of the study area.[14]
Source: Prepared by the author.

The quartzites are interspersed with other metamorphic rocks (metadiabase, metagabbro and metadiacmictite) and, to a lesser extent, with igneous (granite) and metamorphic rocks (quartz schist and phyllites). At the contact between the conglomerates and the argillites and siltstones, metasediments occur. In the valleys of the main rivers there are recent alluvial deposits whose sediments come from upstream.

The areas validated in the field include lithologies with different susceptibilities to denudation and which have repercussions on different landforms. Sorting the rocks in the study area on the scale of susceptibility to denudation established by Crepani et al. (2001), the metasedimentary rocks polarised by quartzites are the most resistant and the sedimentary rocks represented by limestones are the most susceptible to denudation.

Uhlein, Tromprette and Egydio-Silva (1995) present a schematic structural section of the São Francisco Craton and the outer portion of the Araçuaí Belt, which cuts the area of the Corinto and Diamantina Charts in a NW-SE direction. The Bambuí Group rocks located between the Cabral and Espinhaço mountain ranges correspond to horizontal layers that become folded and affected by shear zones, as shown in figure 09. The stratigraphy in the field validation areas is shown in table 08.

---

[14] Enlarged figure included in the Appendix to this dissertation.

33

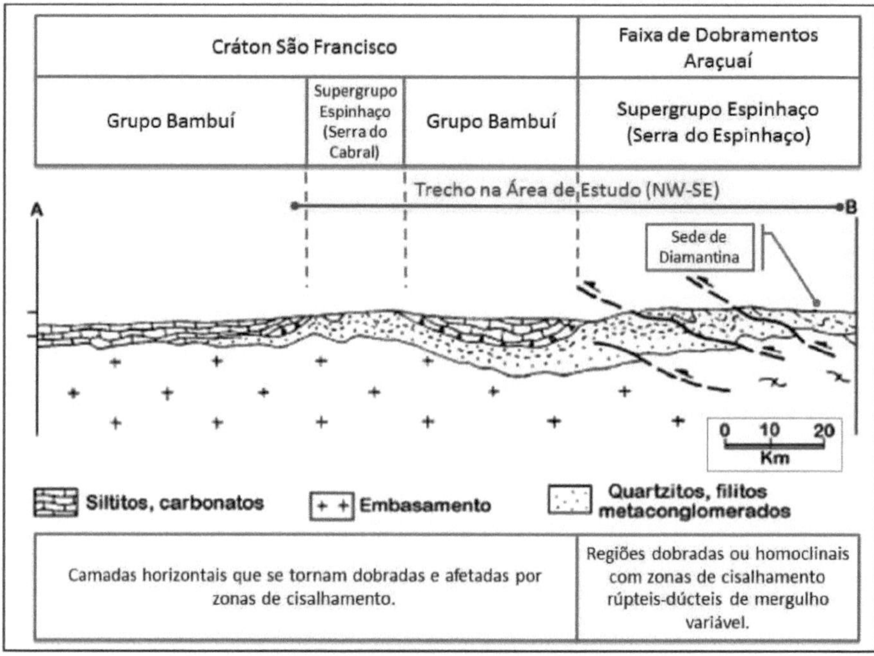

Figure 09 - Lithological section of the study area.
Source: Adapted from Uhlein, Trompette and Egydio-Silva (1995).

Table 08 - Stratigraphy in the field validation areas.

Geochronology	Field validation areas	Lithostratigraphic unit		Lithology
		Supergroup or Group	Training	
Cenozoic	West	Alluvial deposits		Unconsolidated sediments
Neoproterozoic	West and Central	Bambuí	Alligator Lagoon	Limestones and marls
	Central		Serra de Santa Helena	Argillites and siltstones
		Macaúbas Indiviso		Quartzite
Soproterozoic Me	Central and East	Spine	Rio Pardo Grande	Metargillites and metasiltites
	East		Pereira Stream	Quartzite

Source: Prepared by the author based on Uhlein; Trompette; Egydio-Silva (1995), CPRM (2002); Knauer et al. (2011), Guimarães, Travassos and Linke (2011), Fogaça (2012) and Guimarães (2012).

3.2. Geomorphological aspects

The study area has a considerable altimetric range, as shown in figure 10.

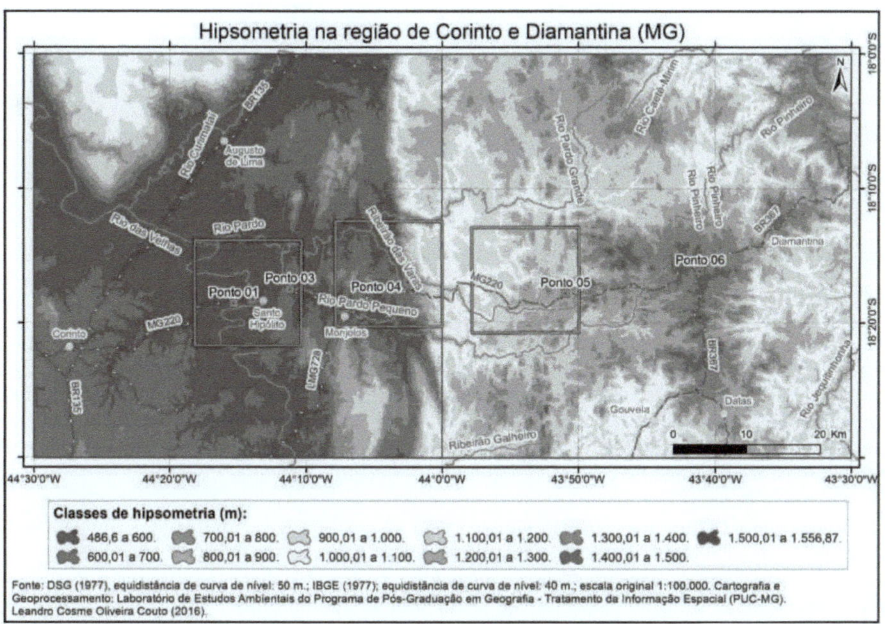

Figure 10 - Hypsometric map of the study area.[15]
Source: Prepared by the author.

The higher elevations of the relief are part of the:

> In Minas Gerais, the Serra do Espinhaço - the great hydrographic dividing line between the basins of central-eastern Brazil and the São Francisco river - is a set of highlands, shaped like a boomerang with a general north-south direction and convexity orientated to the west. The name "mountain range", however, hides a physiographic reality that would be better defined by the term "plateau". (SAADI, 1995, p. 41)

Saadi (1995) also points out that the curved geometry of the Serra do Espinhaço consists of two distinct plateau compartments in terms of lithostructure and morphology: one is southern, with a SSE-NNW direction, and the other northern, with a SSW-NNE direction. They are separated by a depressed zone elongated in a SE-NW direction, located immediately north of Diamantina, in the municipality of Couto de Magalhães de Minas. The lowest altitudes are in the Rio das Velhas plain, which is part of the São Franciscana Depression:

> "(...) a relative depression bounded by the morphostructural domains of the extensive fold belts and associated metasedimentary and Plio-Pleistocene sedimentary coverings to the east and west. To the south it is bordered by units of the crystalline basement, such as the Plateaus of Central-Southern Minas Gerais, and to the north by the northeastern basement, the latter receiving the lower course of the São

---

[15] Enlarged figure included in the Appendix to this dissertation.

The relief unit mapping carried out by the IBGE (2006) identifies the Serra do Espinhaço and the São Franciscana Depression as two distinct geomorphological compartments. While the Serra do Espinhaço stretches from the centre of the state to its northern limit, the Upper Middle São Francisco Depression starts in the centre-west of the state (Serra da Canastra), runs through the northwest and into the interior of the state of Bahia. However, the Institute of Applied Geosciences (IGA), in its general division of the relief of Minas Gerais, distinguishes between these two compartments as:

- *Serra do Espinhaço:* located in the central part of Minas Gerais and developed longitudinally, bordered to the south by the Iron Quadrangle and, when it crosses the northern border of the state, extending into the interior of Bahia; it predominantly shows forms of dissection in rocks that are remnants of ancient planing surfaces that alternate with peaks and ridges elaborated in quartzite exposed in large escarpments generally orientated by fractures.

- *São Franciscana Depression:* located along the São Francisco River drainage, with flattened landforms (around 500 metres above sea level), undulating surfaces and ravines; the landscape in contact with the Serra do Espinhaço shows rugged terrain in which there are hills and ridges with ravines and valleys with the characteristic features of karst landforms developed in carbonate rocks.

The study area covers, to the east, a section of the southern portion of the Serra do Espinhaço and, to the west, a section of the Upper Middle São Franciscana Depression. The difference in altitude along the western slope of the Serra do Espinhaço stands out in the landscape.

Rodrigues (2011) associates the existence of contact between different geomorphological units with lithological differentiations, exemplified by the small but wide valley of the Ribeirão das Varas, near Control Point 04. According to figure 11, the waters of the Ribeirão das Varas drain three different types of terrain:

- The floodplain of the Ribeirão das Varas, settled on quartzites of the Espinhaço Supergroup;

- The slopes of the left bank are steep and made up of limestone outcrops of the Bambuí Group at altitudes ranging from 10 to 20 metres;

- Slopes of the right bank, with strong undulation and composed of metapelites and dolomite lenses belonging to the Espinhaço Supergroup.

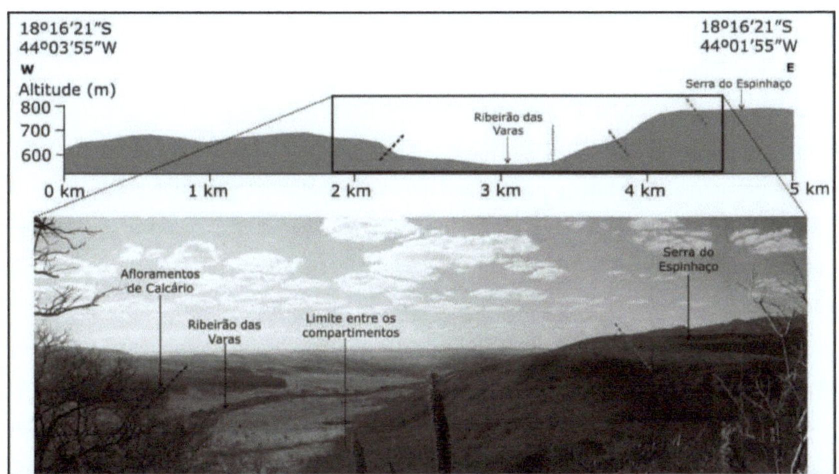

Figure 11 - NNW view of Control Point 04 (18.28° S and 44.04° W) and West-East Altimetric Profile.
Source: Prepared by Rodrigues (2011).

Thus, the lithological contact between the Espinhaço Supergroup and the Bambuí Group manifests itself in the central portion of the study area's landscape through an altimetric break delineated longitudinally from north to south. The Pardo Grande and Pardo Pequeno Rivers (towards the west) and the Pinheiros River (towards the east) rise at altitudes of around 1,400 metres, over quartzite, and flow at their base level at altitudes of around 500 metres. In the western part of the study area, a meandering stretch of the Rio das Velhas gradually consolidates its plain over sedimentary rocks; in the far east, the Rio Jequitinhonha controls the deepening of the drainage network in valleys over quartzite. The existence and dynamics of these valleys corroborate the occurrence of a probable rifting as pointed out by Abreu (1995) and Saadi (1995).

Despite the different origins of the lithological and topographical data, the predominant location of the metasedimentary rocks (e.g. quartzites) correlates with the altitudes

higher altitudes of the landscape (plateau above 900 metres), while sedimentary rocks (e.g. claystones and siltstones) correlate predominantly with the lower altitudes of the landscape (plains below 900 metres). Guimarães (2012) maps these two compartments respectively as Serras, Patamares e Escarpas do Espinhaço and as Superfícies Aplainadas da Depressão Periférica do São Francisco.

However, this general correlation does not apply to sedimentary rocks such as limestones and marls, which are found at altitudes of up to 900 metres in some stretches and are crossed by the Pardo Grande and Pardo Pequeno rivers at altitudes of 600 metres in others. Guimarães, Travassos and Linke (2011, p.239) characterise the karst relief of the Monjolos region "by the presence of massive limestones or low mountains, broad, gentle hills and flattened surfaces characteristic of reliefs built on carbonate rocks with altitudes varying between 520 and 800 metres".

For this region, Guimarães (2012) recognises the existence of flattened surfaces embedded in residual plateaus of the São Franciscan Depression geomorphological compartment. In the study area, the stretch mapped by Guimarães (2012) as the São Francisco Residual Plateaus acts as an altimetric transition between the Serra do Espinhaço plateau and the São Franciscan Depression plain.

In this scenario, the altimetric variations suggest yet another spatial correlation with the lithology, confirmed by the spatial distribution of the slope and topographic roughness of the land model, shown in figures 12, 13 and 14.

Where the slope is above 08 %, implying the classification of the slope of the land model as undulating, strongly undulating, mountainous or steep, there are metasedimentary rocks (mainly) and limestones and marls. There are flat and gently undulating stretches at higher altitudes, making up plateau portions embedded in terrain with a slope of over 08 % (as seen from Control Point 05, whose view is shown in figure 13).

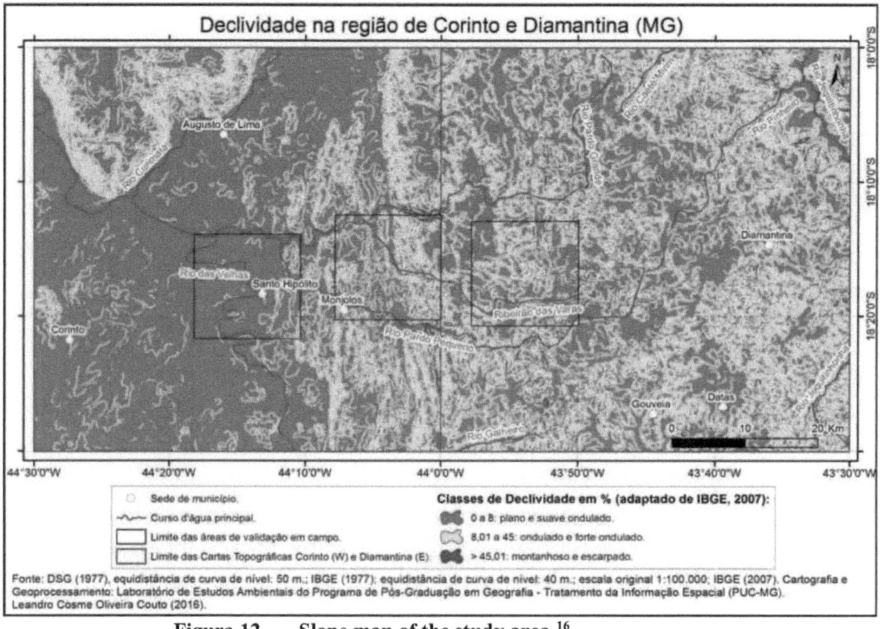

Figura 12   **- Slope map of the study area.**[16]
Source: Prepared by the author.

---

[16] Enlarged figure included in the Appendix to this dissertation.

Figure 13 - ESE view of Control Point 05 (18.29° S and 43.84° W).[17]
Source: Prepared by the author, 19/06/2015.

On the other hand, the other sedimentary rocks occur where the slope is below 08%, implying that the terrain is classified as flat or gently undulating. The predominance of these classes coincides with the lowland portion of the study area's landscape.

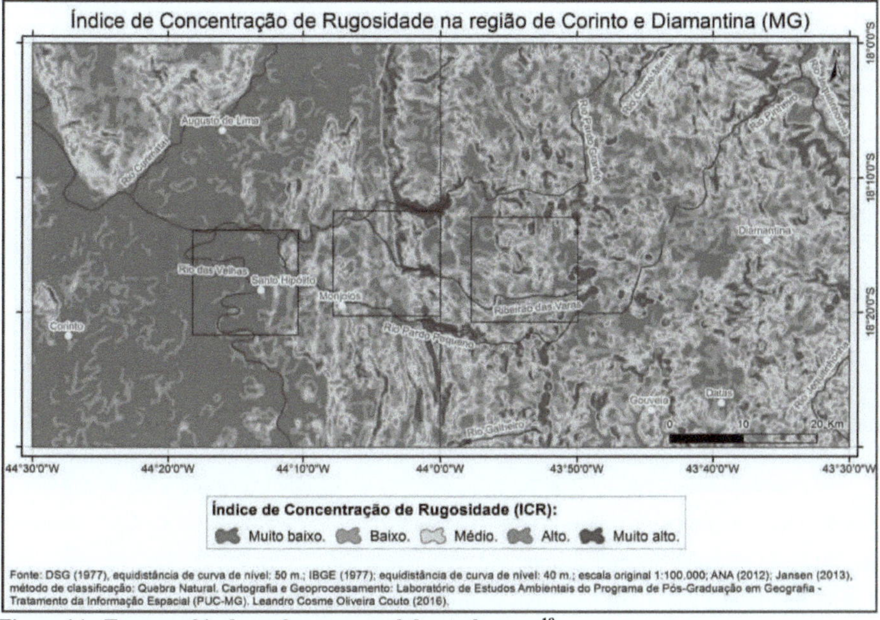

Figure 14 - Topographical roughness map of the study area.[18]
Source: Prepared by the author.

In the topographic roughness map shown in figure 14, it is possible to discern stretches with very low roughness index, both in the largest portion of the plain located to the west in the study area, and in small portions in stretches with high and very high roughness index in the central and eastern portions. Centralised in the study area, the longitudinal alignment with high and very high roughness values attests to the morphostructural contact between the São Franciscana Depression to the west

---

[17] Enlarged figure included in the Appendix to this dissertation.
[18] Enlarged figure included in the Appendix to this dissertation.

and the Serra do Espinhaço to the east. To the north and south of the Monjolos municipal centre, intermediate values of topographic roughness in the Low, Medium and High classes denote the location of karst relief.

2 .3 Natural underground cavities.

The existence of natural underground cavities has a greater potential for occurrence depending on the type of rock. According to the terminology used by Guimarães, Travassos and Linke (2011) and Jansen (2013), among other authors, the speleological conditions that occur in limestones make up the traditional karst landscape and the speleological conditions that occur in quartzites and other rocks make up a non-traditional karst landscape. Table 09 shows the gradation of speleological potential drawn up by Jansen (2011) and validated by CECAV according to lithotype:

Chart 09 - Degree of potential for caves according to lithology.

Lithotype	Degree of speleological potential
Limestone, Dolomite, Evaporite, Metacalcareous, Banded ferriferous formation, Itabirite and Jaspilite	Very high
Calcrete, Carbonatite, Marble and Marga	High
Sandstone, Conglomerate, Phyllite, Shale, Phosphorite, Grauvaca, Metaconglomerate, Metapelite, Metasiltite, Micaxyst, Milonite, Quartzite, Pelite, Rhyolite, Rhythmite, Calci-silicate rock, Siltstone and Shale	Medium
Anorthosite, Arcosseous, Augengnaisse, Basalt, Chamockito, Diabasio, Diamictito, Enderbito, Gabbro, Gnaisse, Granite, Granitoid, Granodiorito, Homfels, Kinzigito, Komatito, Laterite, Metachert, Migmatito, Monzogranito, Oliva gabro, Orthoanfibolito, Sienito, Sienogranito, Tonalito, Trondhjemito, among other lithotypes	Bass
Alluvium, sand, clay, gravel, mudstone, lignite, peat and other sediments	Unlikely Occurrence

Source: Jansen (2011).
Available at:< http://www.icmbio.gov.br/CECAV/projetos-e-atividades/potencialidade-de-ocorrencia- de-cavernas.html>;
Accessed on: 02/2016.

The gradation developed by Jansen (2011) applied to the lithological diversity of the study area generates the map shown in figure 15, which georeferences 75 cavities in the CECAV database (January 2016), 53 of which are in limestone and 22 in siliciclastic rocks.

The limestones and marls stand out in the landscape as having a very high potential for cavities, which is confirmed by the greater occurrence of cavities recorded by the CECAV. In addition, there is a predominance of medium speleological potential in the landscape of the study area, where the other caves recorded by the CECAV occur and where some stretches of low potential or unlikely occurrences are located.

In the study area, limestones and marls make up the traditional, carbonate karst formed by chemical weathering, and are spatially distributed in the municipality of Monjolos and part of Santo Hipólito. The quartzite and other lithologies make up the non-traditional, siliciclastic karst, generated by physical weathering, spatially distributed in the municipality of Diamantina.

Guimarães, Travassos and Linke (2011) describe the exokarst as characterised by limestone

40

massifs and walls covered by lapiás, dolines, poljes, sinkholes and resurgences; the endokarst is better known in the southern part of Monjolos, while the northern part of the municipality lacks more studies.

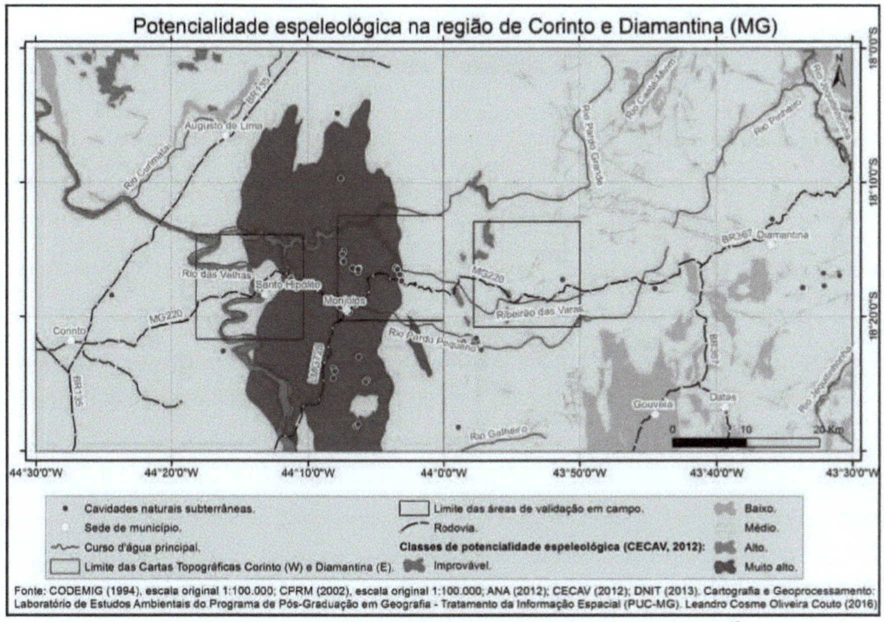

Figure 15 - Map of speleological potential in the study area.[19]
Source: Prepared by the author.

According to Oliveira et al. (2007), cited by Travassos, Guimarães and Varela (2008), the use of caves in the Monjolos region dates back to the 18th century through the extraction of saltpetre (which led to the discovery and degradation of other caves), gradually followed by religious celebrations. Among the natural underground cavities in the traditional karst of Monjolos, Guimarães, Travassos and Linke (2011) highlight the relevant cultural appropriation in 03 cavities:

- Lapa da Fazenda Velha (X: 594,522.315; Y: 7,979,081.769): located in the northern part of the municipality of Monjolos and with an easily accessible entrance through a shelter; it is not very ornate and has two small linear conduits (7 metres long) that indicate a possible past or modern water flow; it is of great speleological importance due to the large number of rock figures distributed on the walls, ceilings, stepped steps and collapsed blocks, with diversity of graphics.

[19] Enlarged figure included in the Appendix to this dissertation.

- Bonina Cave (X: 592,836.474; Y: 7,981.660,138): located in the northern part of the municipality

---

[19] Enlarged figure included in the Appendix to this dissertation.

41

of Monjolos, specifically in the region known as Salobro; it has a linear development of 07 metres, and is a shelter with rock engravings divided into sets of geometric figures on the horizontal supports, zoomorphs and anthropomorphs are located on the vertical wall, similar to the engravings in the Lapa da Fazenda Velha; there are graphisms that record the presence of different visitors, demonstrating the use of the cave in interpersonal and topological relationships (special and resting place).

- Pau Ferro Cave (X: 594,336.304; Y: 7,974,388.071): according to Oliveira et al. (2007), cited by Travassos, Guimarães and Varela (2008) and Guimarães, Travassos and Linke (2011), this is a cave listed as a municipal heritage site, located close to Monjolos, and surrounded by strong religious sentiment in the community; it has five entrances and a skylight and a linear development of 701.8 metres in the preferential longitudinal direction SW-NE; although very ornate with various speleothems, the easy access combined with the lack of guidance for visitors results in the occurrence of damaged speleothems and graffiti on the walls near the entrances; the main conduit is crossed by the Pau Ferro stream, an intermittent tributary of the Rio Pardo Pequeno.

These three caves, among others, are located in the field validation area known as Central, with Gruta do Pau Ferro standing out as the one with the greatest horizontal projection of the three. Specifically, in the West field validation area, only two caves have been identified, both in limestone: Grutinha da Coca and Gruta dos Quatro Morcegos. The other cavities in the Monjolos karst need to be studied and/or researched in order to gain a better understanding, as indicated by Travassos, Guimarães and Varela (2008) and Guimarães, Travassos and Linke (2011).

In the East validation area, on metasedimentary rocks of the Espinhaço Supergroup, only one quartzite cave has been identified, called Lapa do Caboclo, which has been recorded for a long time, according to Travassos, Guimarães and Varela (2008). However, the region also needs to be studied and/or researched in order to gain a better understanding of it, as Fabri (2011) did when he identified 11 quartzite caves in the Itambé do Mato Dentro region.

3.4 Pedological cover

The combination of lithological diversity with the geomorphological aspects of topographic roughness and altimetric differentiation in the study area resulted in a relatively coherent spatial distribution of soil cover, mapped by the Soil Department of the Federal University of Viçosa (DPS / UFV, 2006) and shown in figure 16:

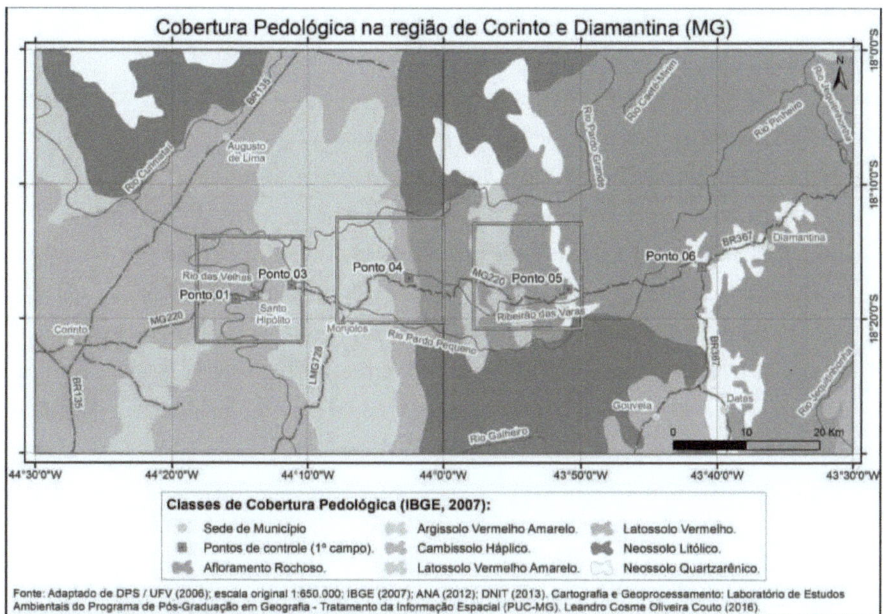

Fonte: Adaptado de DPS / UFV (2006); escala original 1:650.000; IBGE (2007); ANA (2012); DNIT (2013). Cartografia e Geoprocessamento: Laboratório de Estudos Ambientais do Programa de Pós-Graduação em Geografia - Tratamento da Informação Espacial (PUC-MG). Leandro Cosme Oliveira Couto (2016).

Figure 16 - Map of the soil cover in the study area.[20]
Source: Prepared by the author.

Over metasedimentary rocks of the Espinhaço Supergroup there is more recent pedological cover (lithic and quartzarenic neosols) or nonexistent (rocky outcrops); over sedimentary rocks of the Bambuí Group there is older pedological cover (argissolos, latossolos and cambissolos).

The quartzite rock outcrops are interspersed with patches of quartz neosols, which occur azonally over quartz schist, metargillites and metasiltites, in various types of topographic roughness. In the vicinity of the municipal centre of Gouveia, the occurrence of cambissolo háplico is clearly associated with the presence of quartz schist and granite. The occurrence of yellow red latosol amid outcrops of quartzite occurs on metargillites and metasiltites combined with very low topographic roughness.

To the north of the Rio Pardo Grande, to the south of the Rio Pardo Pequeno and around the Serra do Cabral, north-west of the study area, there are lithic neosols predominantly on quartzite associated with an undulating and strongly undulating relief. In the central part of the study area there are argisols over limestones and marls. On claystones, siltstones and conglomerates there are latosols interspersed with cambisols.

Given the generalisation inherent in the small scale of the mapping carried out by DPS / UFV

---

[20] Enlarged figure included in the Appendix to this dissertation.

43

(2006), field verification identified different soil types for the West (Point 01), Central (Point 03) and East (Point 05, complemented by Point 06) validation areas. According to figure 17, in the mapping carried out by DPS / UFV (2006), Point 01 is on cambisolo háplico and close to its contact with red latossolo. However, in the field, it is red latosol. The same occurs at Point 03, which is on clay loam, but in the field is on lithic neosol. Points 05 and 06 are on quartzarenic neosol and close to its contact with a rocky outcrop, but in the field they are on spodosol and plinthosol respectively.

For the East and Central field validation areas, already studied specifically by Guimarães (2012) on a larger cartographic scale, the existence of the types of fluvic neosols, latosols, argissols and cambisols on the Bambuí Group and lithic neosols and rocky outcrops, with patches of latosols on the Espinhaço Supergroup can be seen. Occurrences of vertisols and gleissols were also identified, but due to the scale used for mapping they do not appear on the soil map produced by the author based on the Rio das Velhas Project (CPRM) and shown in figure 18.

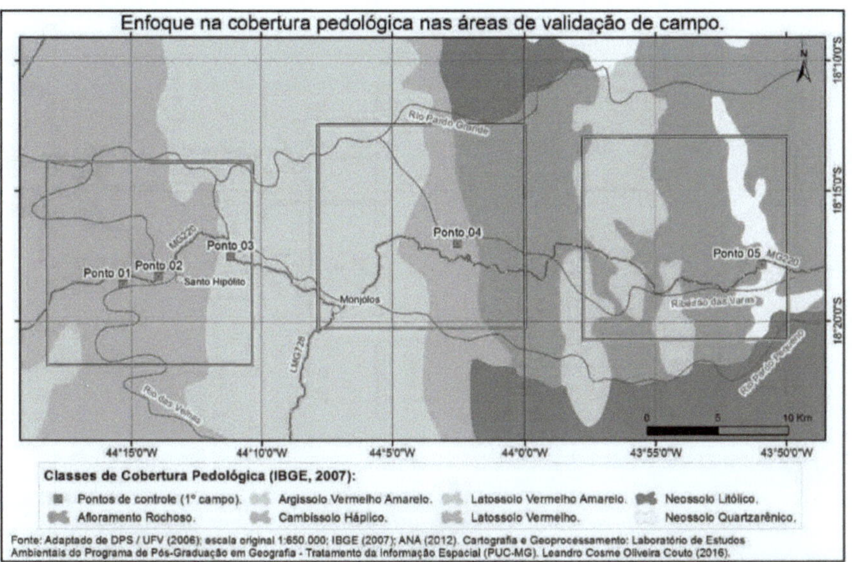

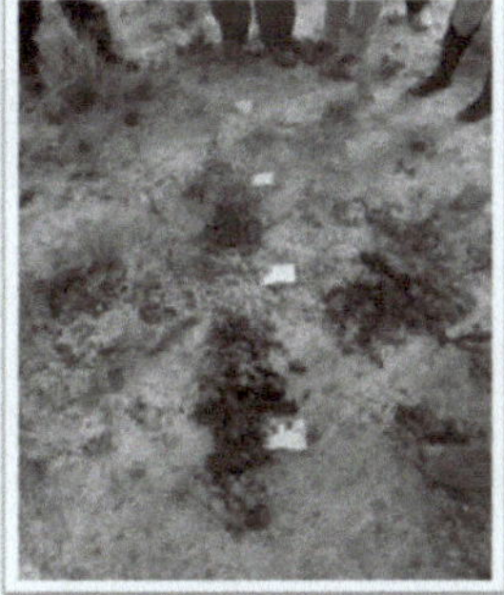

Ponto de controle 01: Perfil de Latossolo.	Ponto de controle 03: Perfil de Neossolo Litólico.	Ponto de controle 05: Tradagem de Espodossolo.

Fonte: Fotos do autor em trabalho de campo realizado de 19 a 21/06/2015 como requisito parcial da disciplina *Estudos Integrados de Meio Ambiente*, ministrada pelo professor Dr. Guilherme Taitson Bueno entre 02/03/2015 e 08/06/2015, integrante do Programa de Pós-Graduação em Geografia – Tratamento da Informação Espacial, da PUC Minas.

Figure 17 - Focus on the field validation areas and records of the soil types identified in the initial fieldwork.

Source: Prepared by the author.

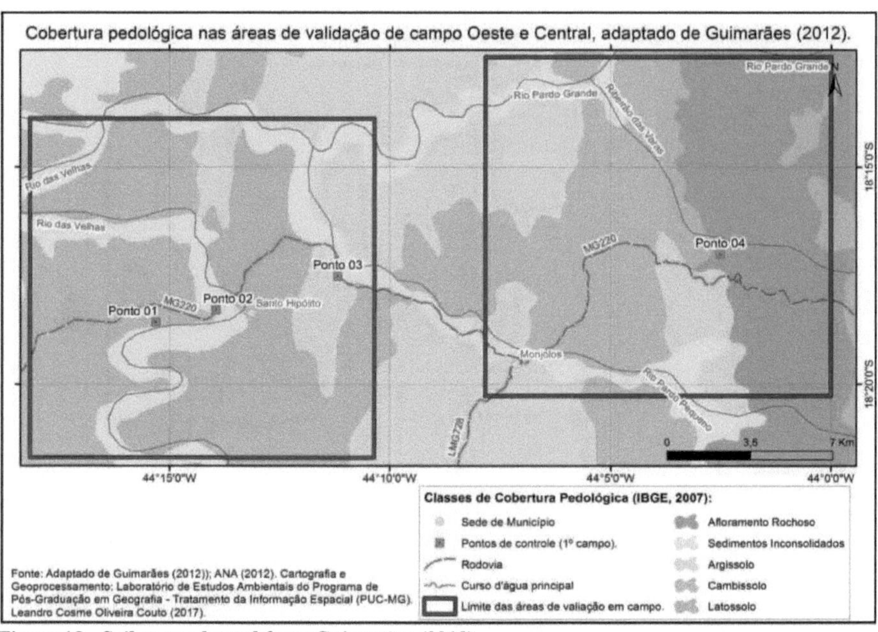

Figure 18 - Soil map adapted from Guimarães (2012).
Source: Adapted from Guimarães (2012).

3.5. Climatic aspects

On the Climate Map of Brazil (IBGE, 1978), the southeastern region has a Tropical Brasil Central zonal climate with various variations (subtypes) associated with topography and/or location. Sant'Anna Neto (2005) distinguishes 14 climatic subtypes in this region, 03 of which exist in the study area (Figure 19).

Similar to the correlation between lithological distribution and altimetric variation, there is also a correlation between climatic distribution and variations in relief. The municipal centres of Santo Hipólito and Monjolos, at low altitudes on the São Franciscana Depression, are subject to a hot climate, with an average of over 18° C all year round, and a semi-humid climate, with 4 to 5 dry months. The seat of Diamantina, at high altitudes on the Serra do Espinhaço, is subject to a sub-warm climate, with an average of between 15° C and 18° C in at least one month of the year, and semi-humid, with 04 to 05 dry months. The higher portions of the relief, on the ridge of the Serra do Espinhaço, are subject to the mesothermal subtype, with an average of between 10° C and 15° C throughout the year, and semihumid, with 04 to 05 dry months.

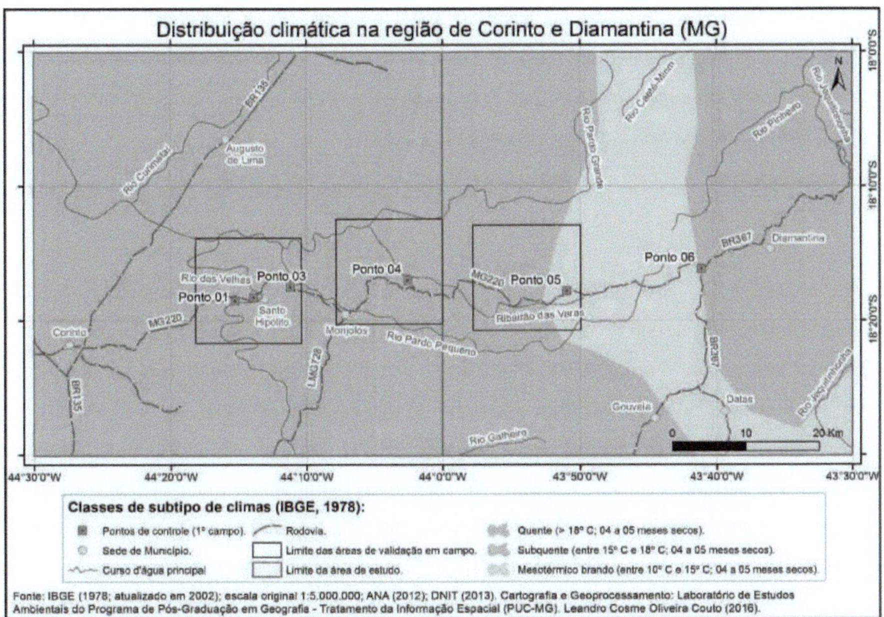

Figure 19 - Map of climate distribution in the study area.
Source: Prepared by the author.

The field validation areas include not only these three subtypes, but also the change from warm to sub-warm climate in the case of the Central validation area, and from sub-warm to mesothermal in the case of the Eastern validation area. Consistent with the climatic conditions of the Central validation area, Guimarães (2012) considers the occurrence of variations for high altitude tropical in mountainous regions of Minas Gerais, such as the Serra do Espinhaço, so that specifically for the study area the climatic characteristics are:

> (...) strongly influenced, among other factors, by the orographic factor. The less elevated areas of the region, corresponding to the municipalities of Monjolos and Santo Hipólito, have a drier and warmer climate, while in the higher altitude regions, corresponding to the Serra do Espinhaço, the climate is wetter and colder. (GUIMARÃES, 2012, p. 101)

Both the modelling proposed by Crepani et al. (2001) and the modelling adapted by Jansen (2013) use the Reduced Index Map (MIR) of the 1:250,000 scale Brazilian Map sheets. On this map, the study area falls within MIR 439. For the National Institute for Space Research (INPE) there are values of 1,280 mm for average annual rainfall and a rainy season of 7.5 months, as well as 170.7 mm of rainfall intensity (the distribution of rainfall in the rainy months).

However, considering the data compiled by Reis, Guimarães and Landau (2012), among the

490 rain gauge stations maintained by the National Water Agency (ANA) and distributed in Minas Gerais, the closest to the study area are those of Curvelo, with an average annual rainfall of 1,225.8 millimetres (mm), and Diamantina, with 1,362.7 mm, whose data corroborates the classification made by the IBGE (1978) and the observation of Guimarães (2012).

3.6. Hydrographic Aspects

The results of the geoecological modelling were validated in the field using three specific areas drained from east to west at different altitudes by the Velhas, Pardo Grande and Pardo Pequeno rivers and the Varas stream. Figure 20 shows a map on which it is possible to discern the altimetric range between the upstream and downstream stretches of these watercourses.

According to the observations of Guimarães (2012), the intense lithological turnover combined with the altitude results in the Serra do Espinhaço being an aquifer recharge area from which numerous springs emerge. The Rio Pardo Grande and its main tributaries (Rio Pardo Pequeno and Ribeirão das Varas) have their sources in the Serra do Espinhaço above 1,300 metres, in the municipality of Diamantina. They flow westwards through the Monjolos karst (also an aquifer recharge area) and flow into the Rio das Velhas, in Santo Hipólito, below 600 metres. All have stretches that coincide with inter-municipal boundaries.

According to Silva, Pedreira and Almeida-Abreu (2005) cited by Guimarães (2012), the drainage pattern in the two basins of the Pardo Grande and Pequeno Rivers is triangular and sub-rectangular with meandering stretches on vast flood plains. Figure 21 shows the drainage network over the topographic roughness in the areas where the models were validated in the field.

Figure 20 - Hypsometry map of the field validation areas,
Source: Prepared by the author.

Combining the field studies of Rodrigues (2011) and Guimarães (2012) with the observations made during the initial fieldwork carried out on 19, 20 and 21/06/2015, the Pardo Grande and Pequeno Rivers and the Ribeirão das Varas, in upstream stretches over the Espinhaço Supergroup, are presented in recessed valleys that take advantage of the intense regional lithological turnover. In downstream stretches over the Monjolos karst there is a relative widening of the valleys of the Rio Pardo Grande, after the mouth of the Ribeirão das Varas, the base level towards the Rio das Velhas.

According to mapping by the DPS / UFV (2006), the Pardo Grande and Pardo Pequeno Rivers and the Ribeirão das Varas drain busy terrain upstream, covered by rocky outcrops and lithic neosols (on rocks of the Espinhaço Supergroup: quartzites, metargilites and metasiltites) and with small stretches of quartz neosols and latosols on flat terrain. On rocks of the Brambuí Group, downstream, these watercourses drain land covered by cambisols and argillites developed on the limestones and marls of the Monjolos karst. From a certain point downstream, these watercourses flow over unconsolidated sediments that they themselves transported from upstream.

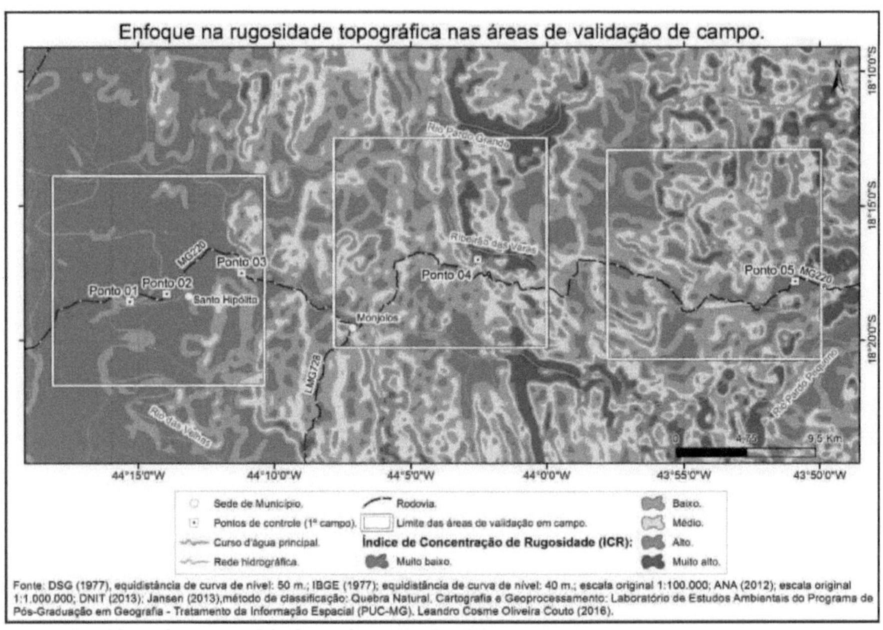

Figure 21 - Map of topographical roughness in the field validation areas.
Source: Prepared by the author.

According to Normative Deliberation (DN) No. 20/1997 of the State Council for Environmental Policy (COPAM), which provides for the classification of waters in the Rio das Velhas basin, the waters of the Rio Pardo Grande sub-basin, from the sources to the confluence with the Rio das Velhas, including the Batatal and Varas streams, the Açougue, Capão and Pindaíba streams and the Rio Pardo Pequeno, are classified as Class 1. By legal definition laid down in National Environmental Council Resolution (CONAMA) No. 357/2005, waters classified in this class are used for human consumption (after simplified treatment), the protection of aquatic communities, primary contact recreation and the irrigation of vegetables and fruit.

3.7. Vegetation cover

Given the geographical position of the study area, native phytophysiognomies typical of the Cerrado Biome predominate, with localised occurrences of phytophysiognomies typical of the Atlantic Forest Biome in stretches of the Diamantina Topographic Chart (E). Figure 22 shows the map of vegetation cover in the study area.

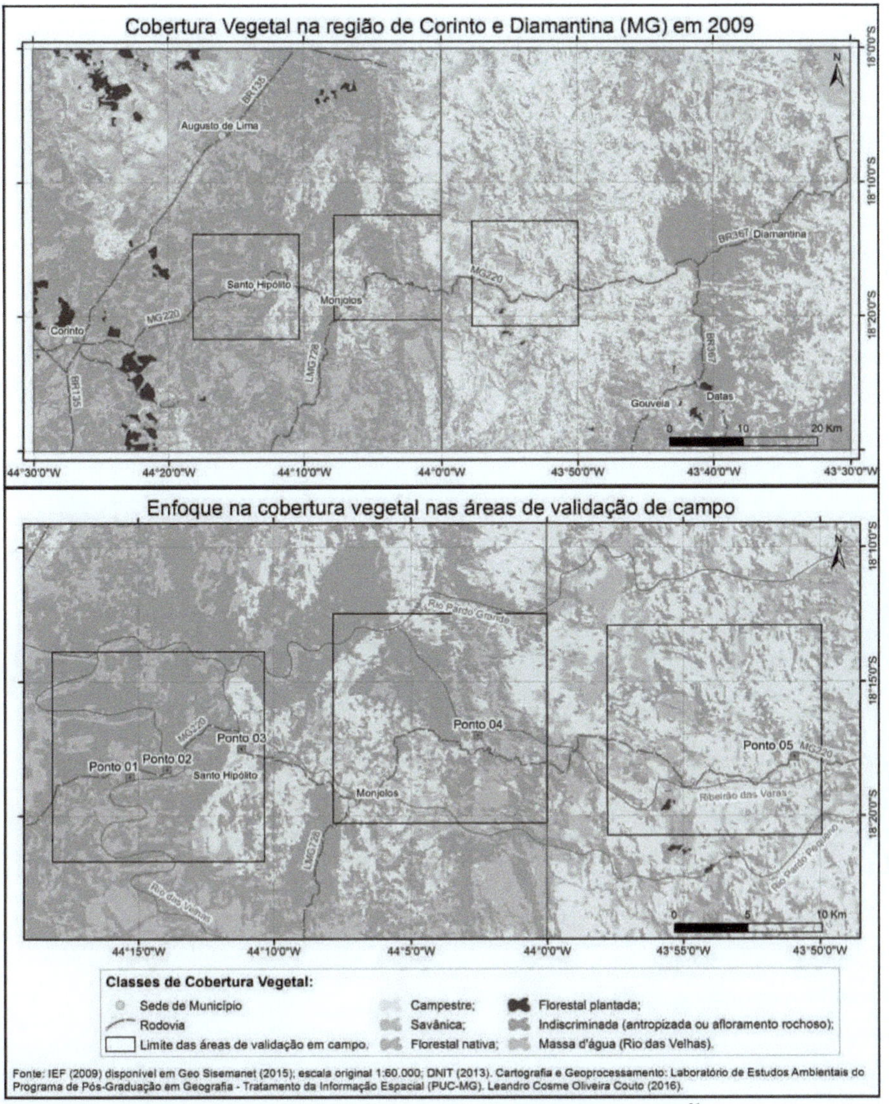

Figure 22 - Maps of vegetation cover in the study and field validation areas.[21]
Source: Prepared by the author.

In order to make the mapping of vegetation cover in the state of Minas Gerais more compatible

General on a scale of 1:65,000 (IEF, 2009), with the other mappings on scales with less detail, the different phytophysiognomies were grouped according to vegetation size as shown in Table 10.

Chart 10 - Correspondence between phytophysiognomies and plant size.

---

[21] *Figures enlarged and included in the Appendix to this dissertation.*

Phytophysiognomy	Plant size
Field and Rupestrian Field	Campestre
Cerrado Grasslands, Cerrado in the Restricted Sense and Veredas	Savannah
Seasonal Deciduous and Semideciduous Forests	Native forest
Eucalyptus or pine forests	Planted forest
Anthropised area or rocky outcrop	Indiscriminate

Source: Adapted from Crepani et al (2001); Embrapa Information Agency.
Available at:
<http://www.agencia.cnptia.embrapa.br/Agencia16/AG01/arvore/AG01 23 911200585232.html>; Accessed on: 03/2016.

In relation to the relief and climatic aspects, savannah vegetation is predominant in the part of the study area in the São Francisco Peripheral Depression, in a hot continental tropical climate and on sedimentary rocks of the Bambuí Group, where there is also anthropisation by farming activities mapped as "indiscriminate" by the IEF. On the other hand, grassland vegetation predominates in the Serra do Espinhaço, in sub-warm continental tropical and mild mesothermal climates over metasedimentary rocks of the Espinhaço Supergroup, as shown in figure 23.

Figure 23 - Grassland vegetation cover at Control Point 05.
Source: Prepared by the author, 19/06/2015.

The validation areas of the field modelling in the West, on the Rio das Velhas plain, and East, on the Serra do Espinhaço plateau, evoke these correlations, as does the Central validation area, on the contact between these geomorphological compartments.

However, the occurrence of karst in the Monjolos region, extending west to Santo Hipólito, breaks this correlation.

According to IEF mapping in 2009, there is both grassland and savannah vegetation cover over the Monjolos karst, as well as anthropised areas. However, Rodrigues (2011) mapped the remnants of the Dry Forest, noting the distribution of forest fragments over the limestones and marls of the Lagoa do Jacaré Formation (Bambuí Group) in association with "a possible structural control developed on the NE-SW axis" (RODRIGUES, 2011, p. 99).

In contrast to the almost complete absence of native deciduous forest cover over the karst region of Monjolos, in the mapping of Minas Gerais carried out by the IEF (2009) by means of remote sensing without validation of the results in the field, Rodrigues (2011) and Rodrigues and Travassos (2013), combining remote sensing techniques and field work, identified considerable areas covered by remnants of dry forest, as shown in figure 24.

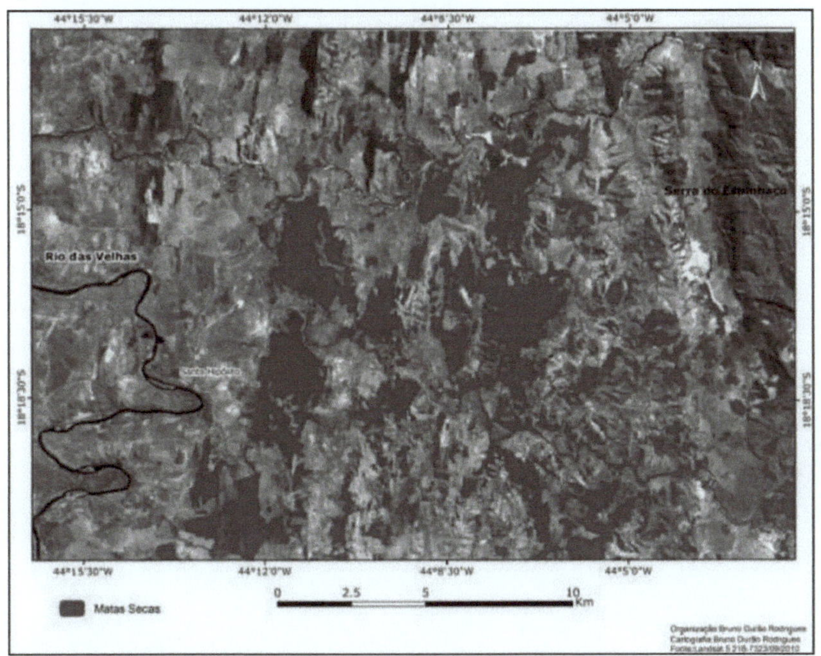

Figure 24 - Remnants of dry forest over the Monjolos karst region.

Source: Rodrigues (2011) and Rodrigues and Travassos (2013).

The occurrence of Dry Forests emphasises the outstanding condition of the Monjolos karst region literally in the midst of the considerable diversity of environmental conditions in the contact between the São Franscisco Peripheral Depression and the Espinhaço Mountains. According to Rodrigues (2011) and Rodrigues and Travassos (2013), the phytophysiognomy of the Dry Forests corresponds to Tropical Dry Forests that are lithodependent on the carbonate substrate and are conditioned by climates that alternate clearly between wet and dry periods. In line with the work of these authors, the field assessment of vegetation cover at Control Point 03, on the banks of the MG 220 and close to the municipal centre of Santo Hipólito, identified the occurrence of native forest vegetation cover (Dry Forest shown in figure 25) instead of savannah (mapped by the IEF). Despite the discrepancies in the mapping of the Dry Forests, mapped as savannah or even grassland cover, the data produced by the IEF (2009) lends itself to the geocological modelling proposed, since the forest phytophysiognomies are noted in the models by Crepani et al. (2001) and Jansen (2013) with more restrictive values than the savannah phytophysiognomies and grasslands.

53

Figure 25 - Remnants of dry forest over the Monjolos karst region, near Control Point 03.

Source: Prepared by the author, 19/06/2015.

CHAPTER 4

GEOECOLOGICAL ANALYSES OF THE LANDSCAPE OF CORINTO AND DIAMANTINA

"Nobody is crazy. Or everyone."
(J. Guimarães Rosa.)

The landscape taxonomy proposed by Bertrand (2004) distinguishes six units (Zones, Domains, Natural Region, Geosystems, Geofacies and Geotopes), among which the Geosystem, with an intermediate spatial scope, stands out for its compatibility with the human scale. The delimitation of these different taxonomic units comes from the identification of objective discontinuities in the landscape, such as the delimitation of different morphoclimatic and phytogeographic domains identified by AB'SABER (2003) for the Brazilian territory, compatible mainly with climatic and geomorphological aspects.

The diversity of geological, geomorphological and climatic aspects in the study area allows the landscape of the Corinto and Diamantina region to be categorised as belonging to the Ecotone that exists between the Cerrado, with tropical plateaus penetrated by gallery forests located in central Brazil, and the forested Mares de Morros, which extend azontally from the north to the south of the country. Thus, the very landscape of the study area, which results from the direct combination of the environmental components of the core areas of these domains, exemplifies an objective discontinuity in the landscape of Brazil's morphoclimatic and phytogeographic domains.

In its western portion, in a stretch covered by the Corinto Topographic Map, it has characteristics closer to those of the Cerrado domain, with climatic conditions typical of the hot continental tropical regime with two distinct rainfall periods and predominantly savannah vegetation. There are enclaves of deciduous forest cover, controlled by the lithology and which differ significantly in the regional landscape.

In its eastern portion, in a stretch covered by the Diamantina Topographic Map, there is grassland vegetation. The high altitudes of the Serra do Espinhaço combine with milder climatic conditions (sub-warm continental tropical and mild mesothermal), with predominantly grassland vegetation with some savannah enclaves. The orography conditions the azonal occurrence of variations in climatic aspects, as well as acting in the azonal control of the occurrence of vegetation. To the east of the Diamantina Topographic Map, outside the study area, there is forest cover.

In this scenario, despite the fact that the study area is located in the ecotone between the Cerrado plateaus and the forested hills, aspects of the Cerrado plateaus domain predominate. Gontijo (2008) states that the Serra do Espinhaço Meridional divides the Cerrado and Atlantic

Rainforest Biomes, so that to the west of the Serra there is a predominance of savannah features and only from the eastern face of the Serra do phytophysiognomies of the Atlantic Rainforest biome occur.

Visualisation of the predominance of the Cerrado domain to the west of the Serra do Espinhaço is identified by making transects from west to east in each of the three areas where the models were validated in the field (West, Central and East), mapped as shown in figure 26. These transects correspond to geosystemic profiles, capable of identifying possible geosystems in the landscape of the study area.

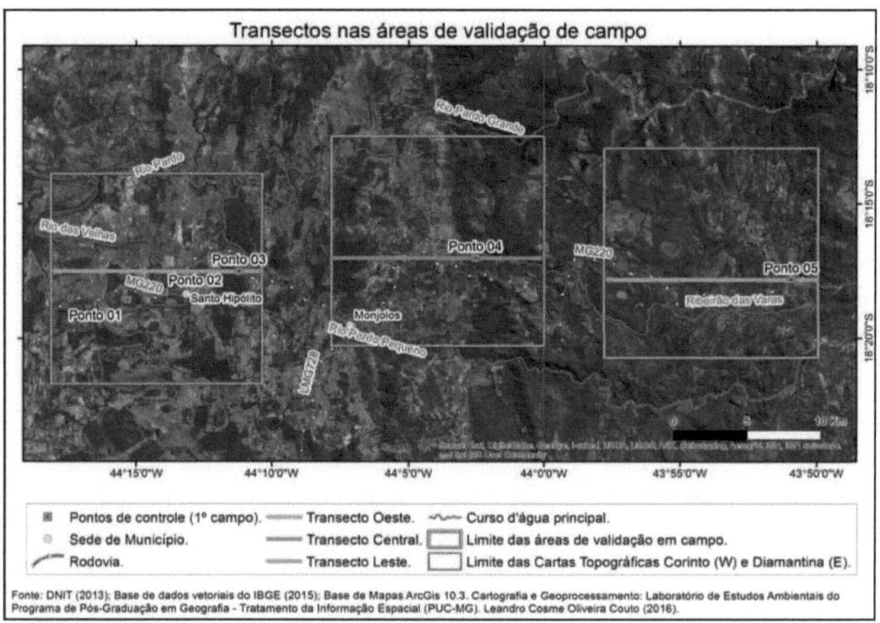

Figure 26 - Map of the transects in the field validation areas.[22]
Source: Prepared by the author.

4.1. Geosystems

Possible geosystems can be recognised by considering them as portions of the landscape with relative ecological continuity, but not necessarily great physiognomic homogeneity, whose transition to other geosystems occurs through ecological discontinuity (BERTRAND, 2004). In this way, the geosystem corresponds to the spatial extension under the same integrated action of climate and lithology, with variations in these aspects as well as in landforms, soil and vegetation cover, and anthropogenic activity. Spatial contact between different geosystems occurs through changes in the aspect of the lithology, even though there are variations in other environmental aspects within the

---

[22] Enlarged figure included in the Appendix to this dissertation.

same geosystem.

Geosystemic profiles, analogous to those proposed by Monteiro (2001) and presented by Jansen (2013), make it possible to visualise the horizontal distribution of the different environmental aspects in the landscape, as well as the vertical overlap of these aspects (as different layers of the landscape). To this end, the data considered is that provided by official bodies (except for vegetation cover) and allows the continuities and discontinuities of the landscape to be visualised in a geosystemic dimension, albeit from the field validation areas as spatial samples.

Specifically to better assess vegetation cover, instead of the data produced by the IEF and referring to 2009, the transects were superimposed on the most recent satellite images of the study area, dated 2014. In addition to the image generated by Landsat 8, we used images made freely available by the free software Google Earth and the map base of the commercial software ArcGIS 10.2, created and updated by the *Environmental Systems Research Institute* (ESRI)[23] .The classification of vegetation cover along the transects was made through visual interpretation of the 03 images combined with the information presented by Rodrigues (2011) and Guimarães (2012) and the observations recorded during the initial fieldwork.

Each transect is tangled with prominent geographical features in the landscape: Rio das Velhas, Rio Pardo Pequeno, Ribeirão das Varas, streams, twice or more along the MG 220 motorway, the urban area of Santo Hipólito, as well as Control Points 03, 04 and 05 for field recording. Figures 27, 28 and 29 show the transects in the West, Central and East field validation areas respectively.

---

[23] This case characterises the fact that the cartographic products generated by official state bodies and/or institutes, which are necessary for studying and monitoring the dynamics of vegetation cover and land use and occupation, are out of date compared to the cartographic products generated by private companies, which are sometimes made available free of charge.

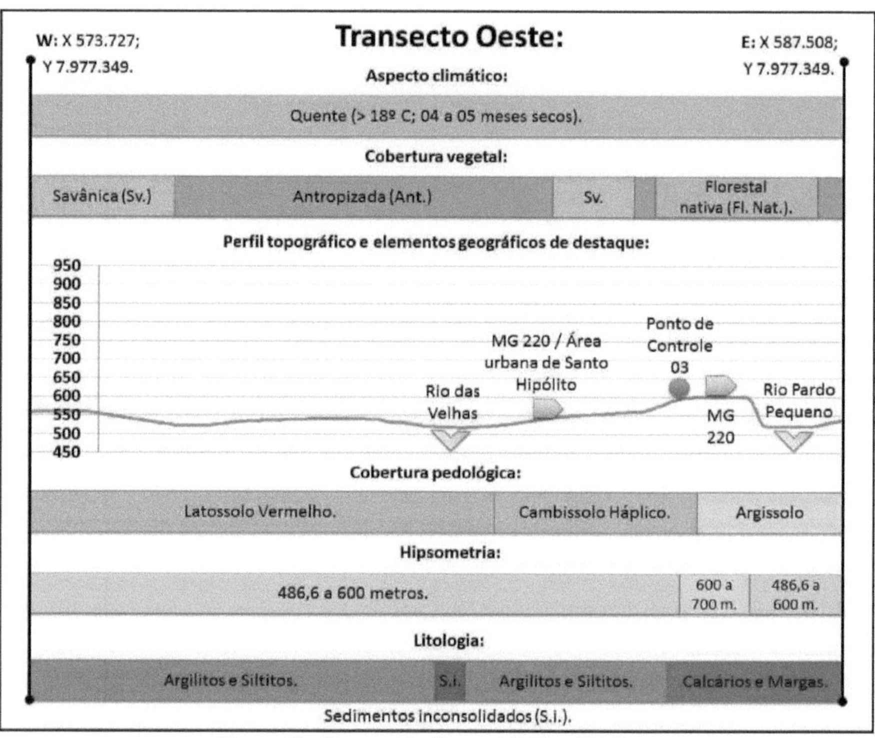

Figure 27 - Transect in the west field validation area.
Source: Prepared by the author.

The Western Transect is developed entirely on rocks of the Bambuí Group and under warm continental tropical climate conditions. However, lithological distinctions have resulted in differences in the geomorphological aspects, as well as in the soil and vegetation cover, along two possible geosystems here called Planície do Rio das Velhas and Carste de Monjolos. Both have the same climatic aspect, but differ notably in their lithological aspect.

The Rio das Velhas Plain geosystem has this river as its local base level and extends predominantly over claystones and siltstones, with occurrences of conglomerates and metargillites and metasiltstones, in terrain with mostly low topographic roughness (flat or gently undulating slopes), at altitudes of less than 700 metres and covered by predominantly aged soils (latosols) with minor occurrences of rejuvenated soils (cambisols). In the midst of the typical lowland landscape there are occasional witness hills, with high to very high topographic roughness at altitudes above 800 metres. The hot continental tropical climate, with wet and dry periods, reflects the occurrence of savannah vegetation, with different phytophysiognomies of the Cerrado Biome, sometimes replaced by anthropisation due to agricultural activities.

Given its location in the hydrographic hierarchy, the Rio das Velhas is also the regional base level, so that the Rio das Velhas Plain geosystem agglutinates a load of physical and chemical sediment from other geosystems and transported by the watercourses. This supply of sediment turns into favourable conditions for agricultural activities.

Through the Rio Pardo Grande and Rio Pardo Pequeno, as well as watercourses and their respective tributaries, especially the Ribeirão das Varas, the geosystems of the Monjolos Karst and the Rio das Velhas Plain receive a load of sediment from upstream.

The Monjolos karst extends over limestones and marls of the Bambuí Group in terrain with topographic roughness varying between very low and very high. This roughness results in an altimetric range of 500 to 800 metres, with undulating and strongly undulating slopes interspersed with flat or gently undulating slopes and covered by rejuvenated soils (mainly argisols, but also cambisols). The annual alternation between wet and dry periods, typical of the continental tropical climate, combined with hot temperatures, is reflected in the predominance of seasonal deciduous forest (Mata Seca), as well as the occurrence of savannah vegetation and anthropogenic agricultural activity.

The Monjolos karst is also identified in the Central Transect which, in addition to the sedimentary rocks of the Bambuí Group, has metasedimentary lithology from the Espinhaço Supergroup and allows the identification of both the contact between these two lithological units and the Monjolos karst geosystem with another geosystem, here called the Serra do Espinhaço Meridional. On the edges of the Monjolos karst geosystem there is rejuvenated pedological cover (cambisols), typical of environments with some degree of physical instability.

Alongside the Central Transect, the Eastern Transect also features the Serra do Espinhaço Meridional geosystem, with both allowing the recognition of two distinct geofacies in this geosystem, which have been named the West Face and the Interfluve.

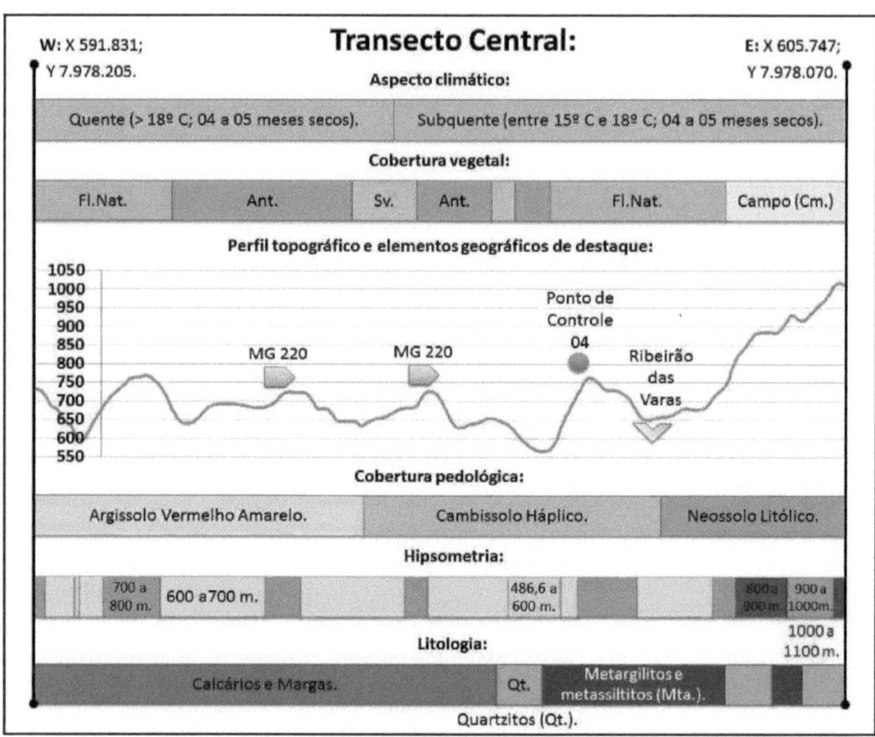

Figure 28 - Transect in the Central field validation area.
Source: Prepared by the author.

The *Serra do Espinhaço Meridional - West Face* geofacies has different climatic and lithological aspects to the *Planície do Rio das Velhas* and *Carste de Monjolos* geosystems. It is under a sub-warm tropical continental climate and extends over metargillites and metasiltites interspersed with quartzites belonging to the Espinhaço Supergroup. This geosystem has the widest altimetric range in the study area, with very low and low topographic roughness terrain, abruptly close to high and very high roughness terrain (flat or gently undulating slopes interspersed with mountainous and steep slopes). In terrain with high or very high topographic roughness, there are rocky outcrops or pedological cover of young soils (litholic neosols) and rejuvenated soils (cambisols). Latosols and quartzarenic neosols are found on land in other topographic roughness classes.

The sub-warm tropical continental climate, combined with the altitude, is reflected in the predominance of grassland vegetation in the form of Campo Limpo, Campo Sujo and, mainly, Campo Rupestre. However, there are also occurrences of savannah vegetation and (little) anthropogenic farming activity. Associated with watercourses there is forest vegetation cover manifested in the Cerradão and Semideciduous Seasonal Forest phytophysiognomies.

The geofacies of the *Serra do Espinhaço Meridional - West Face* is permeated by watercourses that flow predominantly from east to west (Rio Pardo Grande, Rio Pardo Pequeno, as well as their tributaries, especially the Ribeirão das Varas). However, unlike the *Rio das Velhas Plain* and the *Monjolos karst*, the location and topographical conditions of the terrain mean that this geosystem acts as a supplier of physical and chemical sediments. The orogeny of this lithological complex has resulted in a considerable altimetric range, which varies between elevations of just under 600 metres, in the lower reaches of the valleys of the main watercourses, and elevations of over 1,300 metres, in stretches of the upper-middle valley of these watercourses, concentrated between these elevations.

Variations in the climate and lithology recorded in the Eastern Transect, which entirely overlies the metasedimentary rocks of the Espinhaço Supergroup, identify the contact between the *West Face* geofacies and *the Serra do Espinhaço Meridional - Interfluvium*, located at the highest altitudes. This is a geosystem with a mild tropical continental mesothermal climate, which extends mainly over quartzite, metadiabase dykes and metagabbro. To the south and east of the Eastern Transect there are concentrated occurrences of granites and quartz schist. It occupies terrain located from the 1,100 m elevation to over 1,500 m, corresponding to the topographic watershed of the Jequitinhonha River basin to the east and the São Francisco River basin to the west. This geosystem is home to the sources of the region's main watercourses (Rio Pardo Grande, Rio Pardo Pequeno, Ribeirão das Varas) and many tributaries.

The terrain is sloping, with different slopes. In the central part of this geosystem there is a predominance of low and very low topographic roughness indices, associated with flat or gently undulating to wavy slopes. On the edges, the other topographic roughness classes associated with steeply undulating, mountainous and steep slopes predominate. Nevertheless, there are stretches with high topographic roughness in the middle of terrain with low roughness and vice versa.

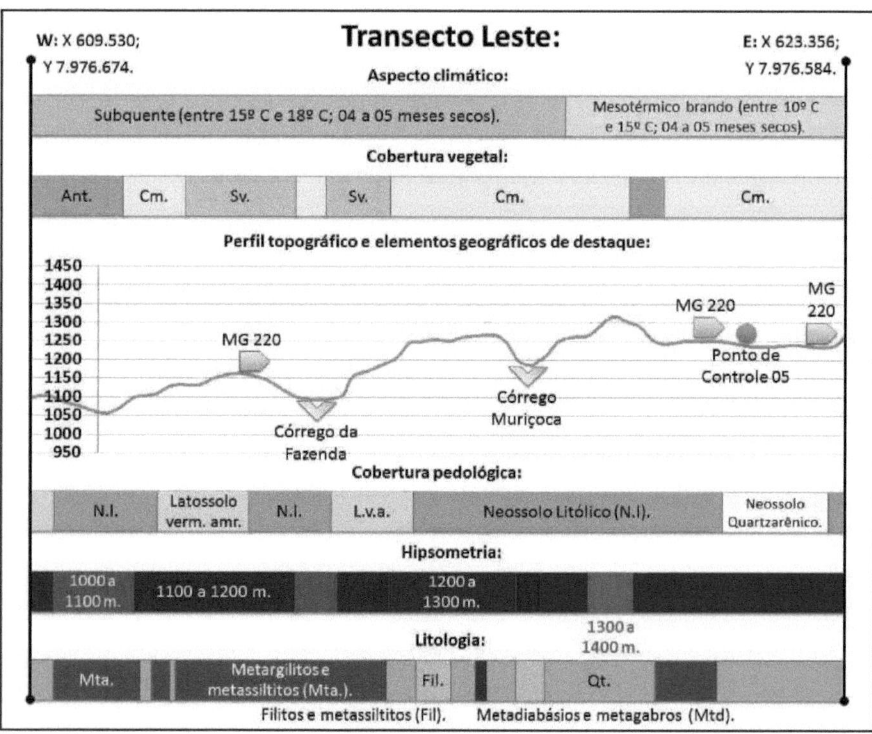

Figure 29 - Transect in the east field validation area.
Source: Prepared by the author.

There is a large occurrence of rocky outcrops interspersed with young soils, with lithic neosols in busy terrain and quartzarenic neosols in less busy stretches. As noted in the field, spodosols also occur in stretches with little movement. The grassland vegetation cover is predominant and the Campo Rupestre and Sujo phytophysiognomies were identified. Little anthropisation was observed.

Given the location of the study area in the Ecotone between the morphoclimatic and phytogeographic domains of the Cerrado (to the west) and the Mares de Morros (to the east), the Rio das Velhas Plain and the Monjolos Karst correspond to landscapes that are part of the Cerrado Domain, while the *Serra do Espinhaço Meridional - West Face* and the *Serra do Espinhaço Meridional - Interfluvio* correspond to differentiated landscapes, typical of the Ecotone, but influenced by this Domain in the case of the *West Face*.

The map of the possible geosystems in the study area is shown in figure 30, the extent of which was inferred by extrapolating the geosystem profiles identified in the three transects and identifying the possible boundaries by checking changes in the lithology. The change in the geosystem from the Rio das Velhas Plain to the Monjolos Karst, and from there to the Serra do

Espinhaço Meridional, does not occur abruptly or simultaneously for all environmental aspects.

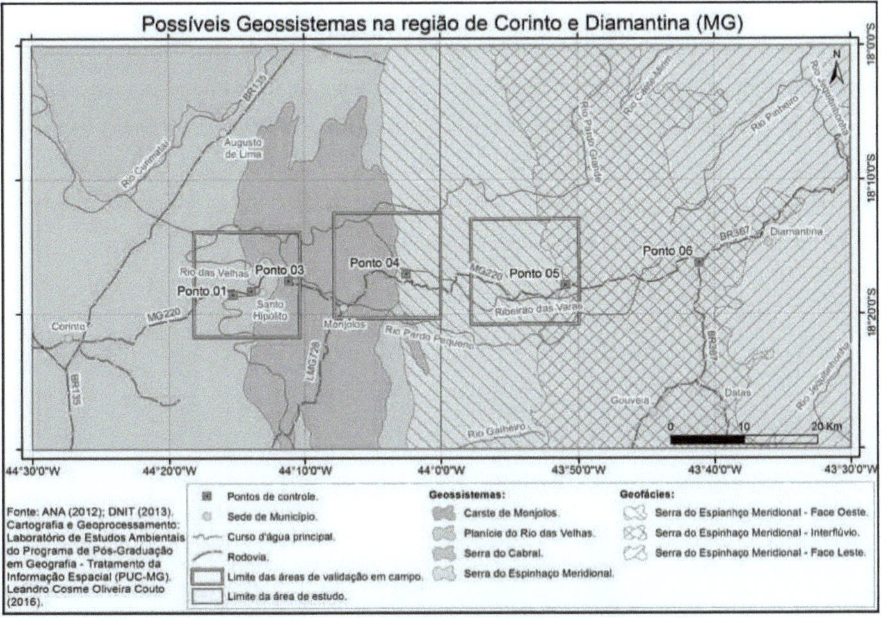

Figure 30 - Map of geosystems in the study area.[24]
Source: Prepared by the author.

In addition to the three possible geosystems (one of which is subdivided into two geofacies), the geosystemic approach to the study area carried out in the office also identified another possible geosystem and another geofacies. These are called the Serra do Cabral geosystem, in the north-west and possibly belonging to the Cerrado Domain, and the Serra do *Espinhaço Meridional - East Face geofacies,* in the east and possibly influenced by the Mares de Morros Domain. The identification of the boundary of the possible *Serra do Cabral* geosystem was inferred by the lithological contact, while the identification of the boundary of the possible *East Face* geofacies was inferred by the change in topographic roughness.

Once the taxonomy of the landscape in the Corinto and Diamantina region has been undertaken through these possible geosystems and their respective geofacies, the application of geoecological modelling based on Tricart (1977), Crepani et al. (2001) and Jansen (2013) allows, respectively, the assessment of morphodynamics, natural vulnerability to soil loss and natural and environmental vulnerabilities in each of these geocological units of the landscape.

---

[24] Enlarged figure included in the Appendix to this dissertation.

4.2. Morphodynamic environments

The assessment of morphodynamic environments in the Corinto and Diamantina region, through the gradation of values of the morphodynamic continuum for the purposes of geoecological modelling of the landscape, identified the occurrence of all seven possible classes. Graph 01 shows the percentage count of the pixels, which indicates the highest occurrence of intergrade medium, followed respectively by the moderate level for stability and instability. Stability and instability also occur at the median level and, in minute quantities, at the strong level.

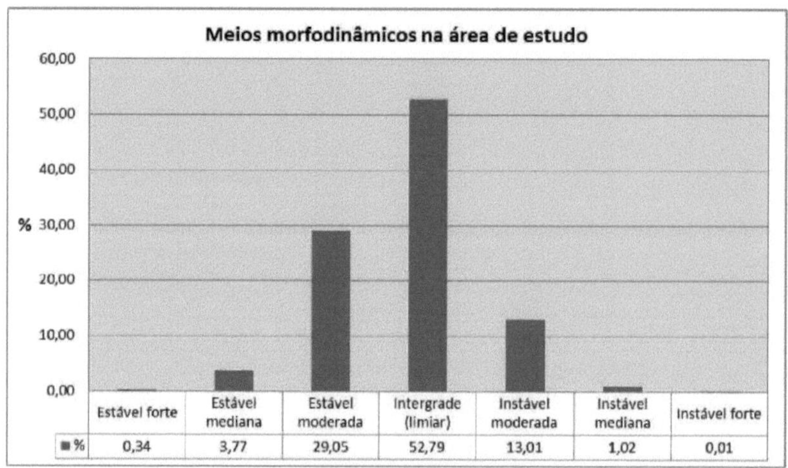

Graph 01 - Percentage distribution of pixels in each class of morphodynamic modelling, based on Tricart (1977), applied to the study area.

Source: Prepared by the author.

However, just as there are differences in quantity between the morphodynamic classes, there are also differences in location, demonstrating a varied spatial distribution. Figure 31 shows the map of the morphodynamic model superimposed on the boundaries of the possible geosystems.

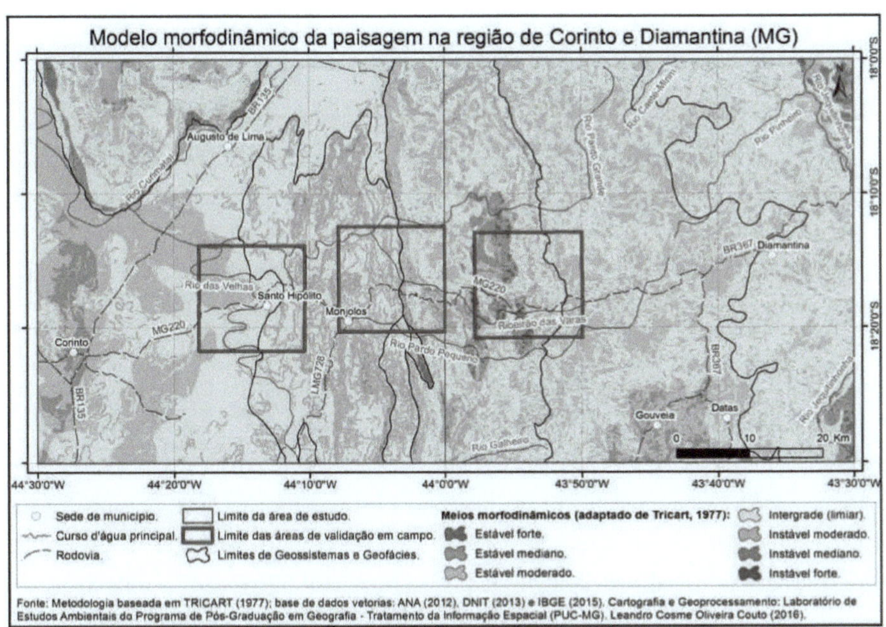

Figure 31 - Map of the morphodynamic model in the study area.[25]
Source: Prepared by the author.

All the possible geosystems identified in the study area have stretches in the intergrade environment, the threshold between stability and instability at the moderate level. According to the analyses of Tricart (1977), in this environment there is permanent competition between pedogenesis and morphogenesis, which exert reciprocal interference. Soils and landforms are of relatively intermediate age, making up landscapes that, when without vegetation cover, are easily susceptible to soil compaction due to cattle trampling. Environmental conservation requires the presence of vegetation cover to inhibit landslides and ravines, so maintaining herbaceous vegetation cover is a minimum condition suitable for agropastoral use, as long as it goes hand in hand with the hydraulic management of runoff to avoid erosion.However, the identification of possible geosystems in the study area allows us to locate some patterns in the spatial distribution (in zones or mosaics) of morphodynamic environments. Table 11 shows the spatial distribution pattern and the predominant morphodynamic medium in each geosystem in the study area.

Table 11 - Spatial distribution and predominant morphodynamic environment in the geosystems of the study area.

Geosystem	Geofacies	Spatial distribution	Predominant morphodynamic environment
Rio das Velhas Plain	-	Zonal	*Intergrade* and moderate stable
Monjolo karst	-	Mosaic	Unstable moderate

---

[25] Enlarged figure included in the Appendix to this dissertation.

Southern Espinhaço Mountain Range	West face	Mosaic	Eclectic, predominantly *intergrade;* possibly unstable
	Interfluve	Mosaic	Stable moderate
	East face	Mosaic	*Intergrade*
Serra do Cabral	-	Mosaic	Eclectic, predominantly *intergrade;* possibly unstable

Source: Prepared by the author.

On sedimentary rocks of the Bambuí Group, the Rio das Velhas Plain is stable, with stretches in intergrade conditions and stable at moderate and medium levels. The Monjolos karst is unstable, with stretches at moderate, medium and strong levels, as well as intergrade stretches.

In line with Tricart's analysis (1977), the morphodynamic stability of the Rio das Velhas Plain is associated with the existence of old soils and landforms. For agropastoral use, it is necessary to specifically study the soil in situ in order to identify improvements and fertilisations. Satisfactory environmental conservation requires maintaining the savannah vegetation cover at a climax density.

The morphodynamic instability of the Monjolos karst is associated with the considerable predominance of morphogenesis over pedogenesis, with the decisive influence of lithological conditions. The relief is rugged, with dissected forms and entombed valleys, and the soils are recent or at least rejuvenated. As well as the exokarst being very active, the endokarst is also active, as this lithology has very high speleological potential. This is an environment at risk of irreversible environmental degradation, which makes agricultural and pastoral uses questionable. At the very least, environmental conservation requires not only the maintenance of vegetation cover, but also hydraulic management of surface runoff. The Monjolos karst has the most unstable morphodynamics in the study area, albeit at a moderate level.

On metasedimentary rocks of the Espinhaço Supergroup, the possible geosystem of the Serra do Espinhaço Meridional has morphodynamic variations in its geofacies. The West Face is the geofacies with the most complex morphodynamics, characterised by a mosaic of intergraded, stable and unstable stretches at a moderate level. There are occasional stretches with medium instability and more expressive stretches with moderate stability.

and strong are observed. The Serra do Espinhaço Interfluve is stable, comprising a predominance of the moderate stable morphodynamic medium, with occurrences at a medium level, and the integrade medium, interspersed with occasional moderate unstable occurrences. The East Face shows a predominance of the integrated environment with occasional occurrences of the stable and unstable environments at a moderate level. Specifically in the valley floor of the Jequitinhonha River, there are significant stretches of stability upstream and instability downstream, at medium and strong levels.

In modelling the West Face, the influence of lithology on unstable morphodynamic environments is clear, with unstable environments occurring at moderate and medium levels on metarhyolites, metasiltstones and phyllites, as well as intergrade and moderate stable environments on quartzites. However, stretches with mountainous or steep slopes were modelled as unstable at the moderate level, or even as moderately stable, rather than unstable at the strong level. This was due, respectively, to the type of lithology and its combination with the type of pedological cover (latossosols).

Given the morphodynamic mosaic, the high altimetric amplitude over a short spatial distance suggests the condition of an unstable environment at the different levels (moderate, medium and strong) on the West Face. There, agropastoral uses are inhibited and environmental conservation is necessary to guarantee the protection of land and watercourses downstream. The *Serra do Cabral* has morphodynamic complexity similar to the *West Face* geofacies, but with more abrupt direct contacts between stable and unstable stretches at moderate and medium levels.

The predominant morphodynamic environment in the *Serra do Espinhaço Meridional - Interfluvio* geofacies is moderate stability, which occurs due to the flat and gently undulating slopes on rocky outcrops of quartzite or lithic neosols. Mid-level stability occurs in localised stretches when associated with rejuvenated soils. *Intergrades* occur occasionally and are the result of undulating and strongly undulating slopes over quartzite rock outcrops or lithic neosols.In an inverse condition to the *Interfluve*, the most common mofordynamic medium in the *East Face* geofacies is the *intergrade*, which occurs due to undulating and strongly undulating slopes over quartzite rock outcrops or lithic or quartzarenic neosols. In this geofacies, the lithology and pedological cover are evenly distributed, with variations only occurring in the slope, conditioning the occurrence of enclaves of moderate stable media (in the case of flat or gently undulating slopes), as well as occasional occurrences of moderate unstable media on mountainous or steep slopes. However, on the right bank of the Jequitinhonha River, variations in lithology (metadiamictites) and pedological cover (latosols) result, respectively, in occurrences of unstable and stable environments, both at medium and strong levels.Taking the integrated set of geofacies of the Serra do Espinhaço Meridional geosystem, the Interfluve has moderate stable morphodynamics, despite its location at high altimetric levels and is subject to the remontant advances of the morphodynamics of the West and East Faces. However, there are differences in the morphodynamics of the West Face, which is more complex and apparently more aggressive towards the Interfluve, and the East Face, which is intergraded and less aggressive towards the Interfluve.In this scenario, map algebra with the 03 variables showed more homogeneous modelled results for the Planície do Rio das Velhas and Carste de Monjolos geosystems, together with the Interfluve and East Face geofacies of the Serra do Espinhaço Meridional geosystem, than for the Serra *do Espinhaço Meridinal - West Face* geofacies and the *Serra do Cabral* geosystem. The indications of the predominant morphodynamic environments in each geosystem, shown in Table 11, as well as the descriptions of their spatial distribution, are complemented by the results of the modelling of natural vulnerability to soil loss proposed by Crepani et al. (2001).

4.3. Natural Vulnerability to Soil Loss

The application of the modelling proposed by Crepani et al. (2001), which results in 21 levels of landscape vulnerability to soil loss units, was carried out by grouping these levels into 5 general degrees of vulnerability, graded between the Stable and Vulnerable classes.

The map algebra used the parameters indicated by Crepani et al. (2001), with the exception of the "Dissection of the Relief by Drainage" due to the incompatibility of the scale of the hydrographic data provided by the ANA (2012), as well as the lack of standards in the data provided by the DGS (1977) and IBGE (1977). The "Altimetric Amplitude" parameter was calculated using the boundaries of the Ottocoded Hydrographic Basins at level 06 by the ANA (2012), the only level that proved compatible with the size of the study area. Figure 32 shows the map containing the result of modelling natural vulnerability to soil loss, applied to the landscape of the study area.

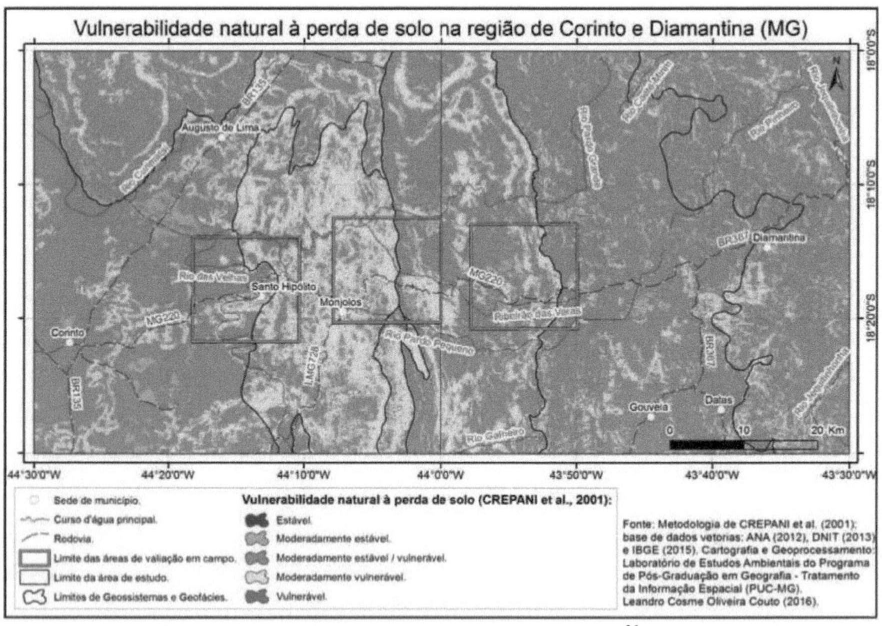

Figure 32 - Map of natural vulnerability to soil loss in the study area. [26]
Source: Prepared by the author.

Since the "Altimetric Amplitude" and "Erosivity" parameters (derived from the rainfall index) vary little or not at all over the study area, the differences in the degree of vulnerability arose due to the spatial distribution of the lithology, slope, maturity of the pedological cover and the density of the vegetation cover. As mentioned above, although the vegetation cover data provided by the IEF

---

[26] Enlarged figure included in the Appendix to this dissertation.

(2009) requires revision for the areas where Dry Woodlands occur, its use for modelling purposes registers vegetation cover density with high values of susceptibility to soil loss in the Monjolos karst geosystem. This distortion ends up emphasising the fragile condition of this geosystem, but does not disqualify the use of the data, since high values of susceptibility to soil loss occur significantly in the other geosystems identified.

In general, the Moderately stable / vulnerable class predominates in the study area, occurring in all the possible geosystems identified. This is followed by the Moderately stable and Moderately vulnerable classes, and there are no records of the stable or vulnerable classes. Graph 02 shows the percentage count of pixels in each vulnerability class from the Crepani et al. (2001) model.

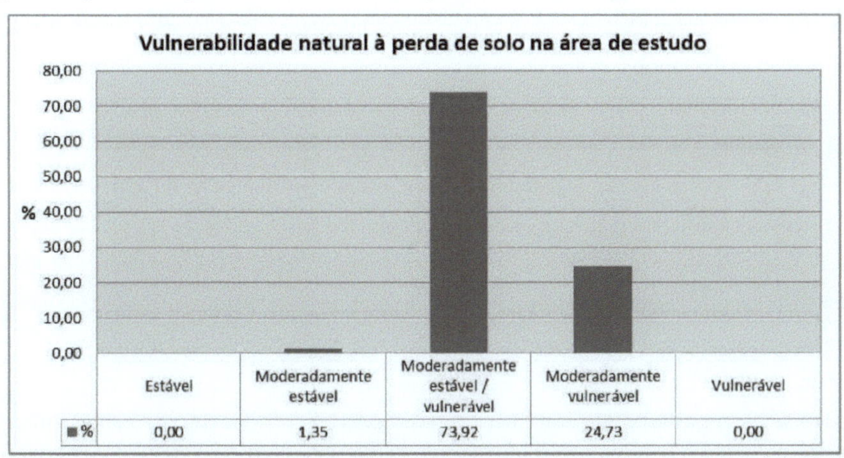

Graph 02 - Percentage distribution of pixels in the natural vulnerability to soil loss modelling classes, based on Crepani et al. (2001), applied to the study area.
Source: Prepared by the author.

Cases of application of the model by Crepani et al. (2001) in other areas (sub-basins of the Alto Rio Pardo - SP, the Alto Rio das Velhas - MG, the municipality of Morretes - PR) show similarity in the pattern of results, with the Moderately stable / vulnerable class predominating with a very high percentage, followed by the Moderately stable class or the Moderately vulnerable class with lower percentages, and the Stable and Vulnerable classes with significantly lower percentages (RIBEIRO; CAMPOS, 2007; SANTOS; SOBREIRA, 2008; MESQUITA; ASSIS; SOUZA, 2010).

Once again proving to be the most fragile geosystem in the study area, in the karst of Monjolos the moderately vulnerable class predominates, associated with rocks more susceptible to chemical weathering (limestones and marls), and rugged slopes (undulating, strongly undulating, mountainous or steep). In the Rio das Velhas Plain, as well as in the *Serra do Espinhaço Meridional - West Face,* there are moderately stable stretches, associated with latosols, as well as moderately vulnerable ones, associated with steep slopes (in the case of the *Plain*) or with metarhyolites, metasiltites and phyllites (in the case of the *West Face*). The results from the *West Face*, again similar to the *Serra do* Cabral, demonstrate that it is the most complex geofacies in the *Serra do Espinhaço Meridional*. In the geofacies

Interfluve and on the East Face the moderately stable / vulnerable intermediate class predominates, largely associated with the occurrence of quartzite rock outcrops (absence of soil).

The moderate vulnerability of the Monjolos karst as a result of the application of natural vulnerability modelling to soil loss suggests that environmental aspects act together, with lithology playing a decisive role. The hierarchy of lithological susceptibility to chemical denudation (CREPANI et al., 2001) for modelling already differentiates lithologies. Rocks composed of calcite and dolomite (such as the limestones and marls of the Monjolos karst) are the most susceptible and rocks composed of siliciclastic miners (such as the quartzites of the Serra do Espinhaço Meridional) are the least susceptible.

Saadi (1995) also recognises that the lithological complex of the Serra do Espinhaço and surrounding areas has a different influence on the evolution of the landscape in the region due to the differences in their behaviour in the face of weathering. Among the lithostructural units that this author distinguishes for the region's landscape, in the study area, are the predominantly quartzite units of the Espinhaço Supergroup, cut by dykes of metabasic rocks and the pelitic-carbonate unit of the Bambuí Group.

In karst landscapes, the pedological cover interacts decisively with the epicarst (upper portion of the underlying soluble rock covered by unconsolidated material or not), "influencing the internal water circulation and the elaboration of the covered rock morphology, as well as the exocarst and endocarst" (PILÓ, 2000, 94). The occurrence of rejuvenated soils (argisols and cambisols) indicates evolution in a humid environment, with the agility of chemical weathering on the impure limestones (marls). This is due to the rapid dissolution of calcite, combined with the rapid leaching allowed by the intense cracking of this type of rock. In this way, the clay material resulting from weathering comes from the impurities, as shown in figure 33.

Figure 33 - Margous limestone (impure) on the banks of the MG 220 motorway, near Control Point 03.

In the siliciclastic landscape developed on quartzite, the effects of geochemical denudation in the lithology of the Serra do Espinhaço Meridional, measured by Rocha (2011), indicate the occurrence of Quartzarenic Neosols or Litholic Neosols, showing a tendency for the soil's sand content to be very high and the clay content very low in lithologies of rocks predominantly composed of quartz. This is due to the difficult chemical weathering of quartz minerals. Although this condition points to an understanding of greater vulnerability to soil loss, the application of the modelling by Crepani et al. (2001) showed predominantly moderately stable / vulnerable results in the Serra do Espinhaço Meridional geosystem, with moderately vulnerable enclaves especially on the West Face.

The simple map algebra, with the 06 variables, shows a landscape classified as moderately vulnerable, despite the fact that the modelling proposed by Crepani et al. (2001) shows some tendency towards intermediate results on the scale of vulnerability to soil loss. On this scale, the most comprehensive class is exactly the moderately stable / vulnerable intermediate class, which should measure the distinction between stability and vulnerability. Based on the Crepani et al. (2001) model, Jansen (2013) developed a more complex map algebra modelling system that proved capable of discerning Environmental Vulnerability in different parts of the landscape, such as the different geosites and geofacies identified in the study area.

4.4. Natural and Environmental Vulnerabilities

The geoecological modelling proposed by Jansen (2013) takes place in two distinct stages, the results of which are, respectively, a Natural Vulnerability model and an Environmental Vulnerability model, both of which are highly compatible with the geosystemic approach proposed by Bertrand (2004). The first, resulting from the weighted map algebra of 05 environmental variables, refers to the ecological potential of the landscape in Bertrand's terms. The second, the result of simple map algebra between the Natural Vulnerability model and the vegetation cover variable, refers to both the biological exploitation of the landscape and anthropogenic action.

Graph 03 shows the distribution of pixels in the Natural Vulnerability classes with the highest percentage values classified as low, medium and very low in the area, followed by the high and very high classes. Counting this percentage indicates the predominance of higher, more resistant ecological potential in the landscape of the study area.

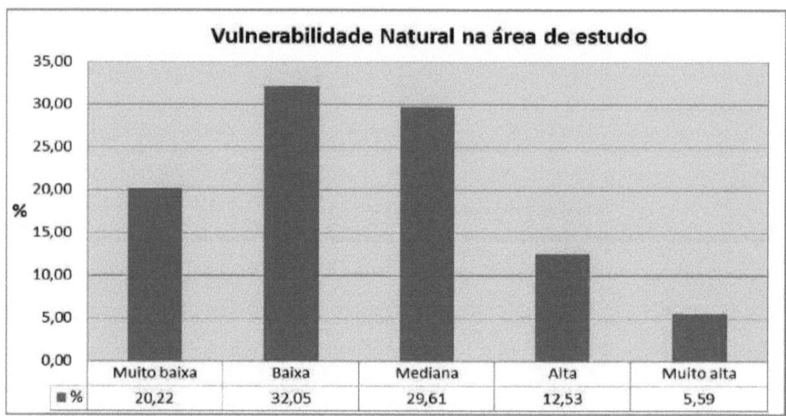

Graph 03 - Percentage distribution of pixels in the Natural Vulnerability modelling classes, based on Jansen (2013), applied to the study area.

Source: Prepared by the author.

Figure 34 shows the spatial distribution of Natural Vulnerability in the study area. There is a concentrated zonal occurrence of very high Natural Vulnerability in the Monjolos karst geosystem, as well as local occurrences in the other geosystems, with emphasis on the West Face geofacies of the Serra do Espinhaço Meridional and the Serra do Cabral. In the Rio das Velhas Plain geosystem, sedimentary lithology conditions medium vulnerability, especially around the Monjolos karst, while occurrences of aged soils (latosols) support very low vulnerability.

In the geosystems of the Serra do Espinhaço Meridional and Serra do Cabral, Natural Vulnerability predominates in the low class, reflecting the preponderance of resistance to chemical weathering of the metasedimentary lithologies. Enclaves of Vulnerability in the very low and medium classes occur reflecting respective variations in topographic roughness and soils: lower roughness and/or aged soils lead to very low natural vulnerability, while high roughness and/or young soils lead to medium vulnerability.

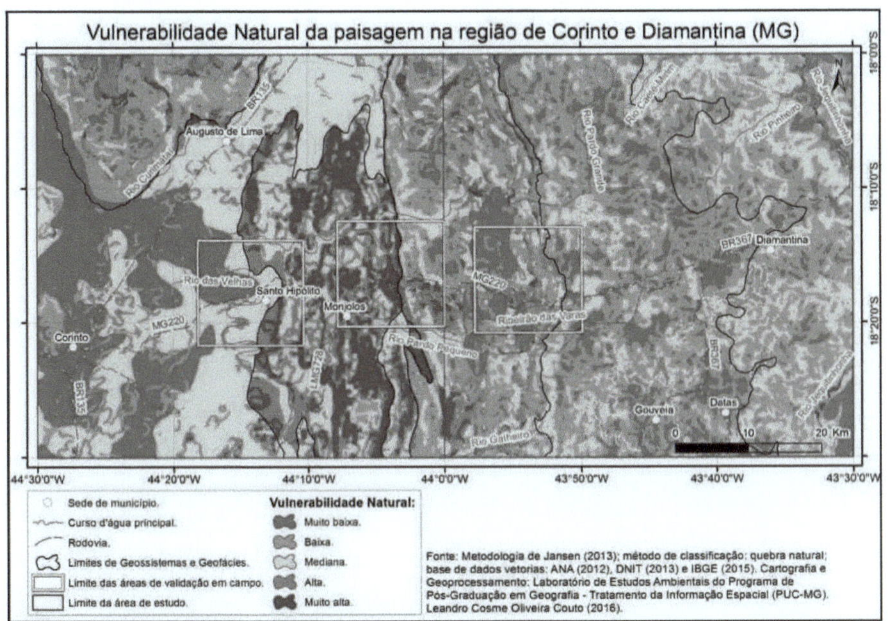

Figure 34 - Map of Natural Vulnerability in the study area.[27]
Source: Prepared by the author.

Paying attention to distinctions in the geofacies of the Serra do Espinhaço Meridional, the Interfluve tends towards very low Vulnerability, while the East Face tends towards medium Vulnerability; the West Face, similar to the Serra do Cabral geosystem, has enclaves of high and very high Vulnerability associated with sandy soils in areas of greater slope. Table 12 shows the predominant Natural Vulnerability in each geosystem in the study area.

Table 12 - Spatial distribution and predominant Natural Vulnerability in the geosystems of the study area.

Geosystem	Geofacies	Spatial distribution	Predominant natural vulnerability
Rio das Velhas Plain	-	Zonal	Average and very low
Monjolo karst	-	Mosaic	Very high
Southern Espinhaço Mountain Range	West face	Mosaic	Low-key, eclectic
	Interfluve	Mosaic	Low, very low trend
	East face	Mosaic	Low, medium trend
Serra do Cabral	-	Mosaic	Low

Source: Prepared by the author.

The Natural Vulnerability model combined with the vegetation cover variable resulted in

---

[27] Enlarged figure included in the Appendix to this dissertation.

modelling the Environmental Vulnerability of the landscape in the study area. Reflecting the equal consideration of ecological potential and biological exploitation, the percentage of pixels in each Environmental Vulnerability class, shown in Graph 04, differs greatly from the percentage in the Natural Vulnerability model.

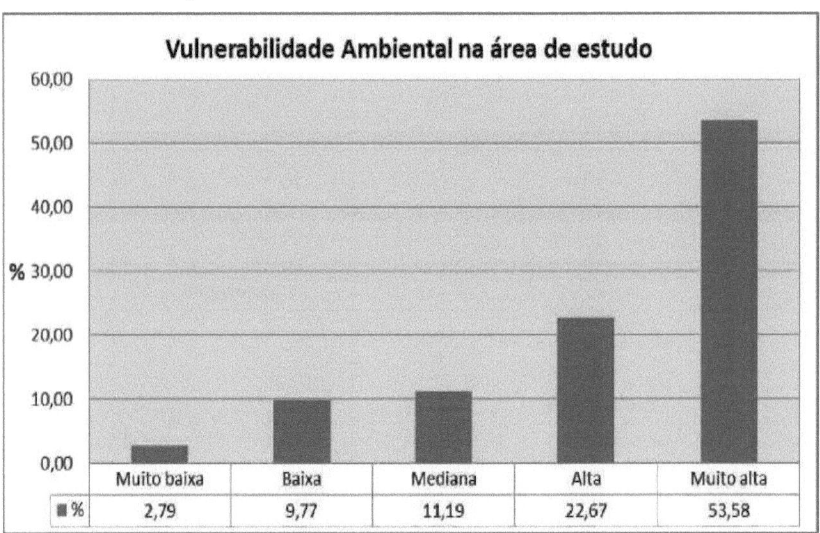

Graph 04 - Percentage distribution of pixels in the Environmental Vulnerability modelling classes, based on Jansen (2013), applied to the study area.

Source: Prepared by the author.

In an exponential distribution of percentages, the study area shows very high Environmental Vulnerability in more than half of the landscape, with the other classes predominating from high to very low respectively. Figure 35 shows the map of Environmental Vulnerability in the study area and makes it possible to identify the variation in spatial distribution in each geosystem.

Figure 35 - Environmental Vulnerability Map of the study area.[28]
Source: Prepared by the author.

In the Rio das Velhas Plain and Monjolos karst, very high environmental vulnerability due to anthropogenic action predominates, with enclaves of biological exploitation (native cover) in the middle class throughout the Monjolos karst, as well as in the middle class in the eastern portion of the Rio das Velhas Plain and low and very low class in the western portion.

Among the geofacies of the Serra do Espinhaço Meridional, high Environmental Vulnerability predominates on the West Face and in the Interfluve, with a greater number of enclaves in the very high class, followed by the low and medium classes, although there are occurrences of the medium class on the West Face and the low class in the Interfluve. In contrast, Environmental Vulnerability is predominantly very high on the East Face, with enclaves mainly in the high class, in addition to the other classes. Serra do Cabral stands out for its considerable eclecticism of Vulnerability, with no predominantly very high classes. Table 13 shows the predominant Environmental Vulnerability in each geosystem.

Chart 13 - Spatial distribution and predominant Environmental Vulnerability in the geosystems of the study area.

Geosystem	Geofacies	Spatial distribution	Predominant environmental vulnerability
Rio das Velhas Plain	-	Mosaic	High

---

[28] Enlarged figure included in the Appendix to this dissertation.

Monjolo karst	-	Mosaic	Very high
Southern Espinhaço Mountain Range	West face	Mosaic	High
	Interfluve	Mosaic	High
	East face	Mosaic	Very high
Serra do Cabral	-	Mosaic	Eclectic

4.5. Integrated analysis of the models

The results of the modelling based on Tricart (1977) and the modelling proposed by Crepani et al. (2001) and Jansen (2013) allow for improved identification, according to Bertrand (2004), of the possible geosystems in the landscape of the study area. Each model emphasises specific geoecological characteristics in its own way, so that integrating the results better individualises and characterises the geosystems identified. The geosystem can be understood as a category of landscape analysis that incorporates the guiding principles of Classical or Traditional Geography listed by Amorim Filho (2006), ensuring that the application of each modelling is better integrated with the others. Table 14 brings together the modelled results, based on Tricart (1977), Crepani et al. (2001) and Jansen (2013), predominant in each geosystem.

Table 14 - Predominant modelled results in the geosystems of the study area.

Geosystem	Geofacies	Morphodynamic environment	Natural vulnerability to soil loss	Natural Vulnerability	Environmental V ulnerability
Rio das Velhas Plain	-	*Intergrade* and moderate stable	Eclectic, Moderately stable / vulnerable	Average and very low	Eclectic, mostly very tall
Monjolo Castle	-	Unstable moderate	Moderately vulnerable	Very high	Very high
Southern Espinhaço Mountain Range	West face	Eclectic, *intergrade* predominance	Eclectic, Moderately stable / vulnerable majority	Eclectic	High
	Interfluve	Stable moderate	Moderately stable / vulnerable	Low	High
	East face	*Intergrade*	Moderately stable / vulnerable	Low and medium	Very high
Serra do Cabral	-	Eclectic, *intergrade* predominance	Moderately stable / vulnerable	Low	Eclectic

The Rio das Velhas Plain is a geosystem whose physical characteristics, typical of its ecological potential, are environmentally resistant. Pedogenesis predominates in the morphodynamics and is combined with the moderate stability of the pedological cover produced, generating aged soils (latosols) on a topography that is not very rugged, generally favouring anthropic action. Even so, there are occurrences of rejuvenated soils (cambisols), which are more fragile and do not favour anthropogenic action, such as the stretch shown in figure 36, which has been subjected to intense laminar erosion. Biological exploitation, consisting of native savannah vegetation cover, generates stretches with low or medium Environmental Vulnerability, while stretches with anthropogenic action consisting of planted eucalyptus forests were modelled with very low Environmental Vulnerability, since the modelling proposed by Jansen (2013) does not distinguish between native forests and planted forests, only recognising which forest parameter is more resistant to savannah and grassland.

Figure 36 - ENE view from coordinates 18.36° S and 44.27° W: area anthropised by abandoned pasture on cambisols on gently undulating terrain.

Source: Prepared by the author.

Immediately to the west of the Rio das Velhas Plain is the Monjolos karst, the most environmentally fragile geosystem in the landscape of the study area and the one with the smallest spatial extent. Its fragility stems directly from its ecological potential, whose limestone and marl lithology is the most susceptible to chemical weathering, resulting in eclectic topographical ruggedness, in which terrain with gently undulating and steep slopes are abruptly close together, as can be seen in figure 37. Morphogenesis predominates in the morphodynamics of the landscape, with moderate natural vulnerability to soil loss.

Figure 37 - NNW view from coordinates 18.28° S and 44.04° W: gallery forest of the Ribeirão das

77

Varas and native grassland vegetation cover on the Margoso Limestone massif bordered by escarpments.

Ecological exploration is also diverse in size, with native vegetation of the grassland, savannah and forest types, the latter being directly associated with the carbonates (Matas Secas). The predominance of very high Environmental Vulnerability occurs both in stretches impacted by anthropogenic action and in stretches with grassland vegetation cover. The savannah cover results in enclaves modelled with medium Environmental Vulnerability. If the IEF database discriminated against the occurrence of Dry Forests, as identified by Rodrigues (2011) and Rodrigues and Travassos (2013), enclaves with medium or even low Vulnerability would be modelled.

The difference in ecological potential and ecological extraction between the Monjolos karst and the Rio das Velhas plain is so marked that it cannot be recognised as a geofacies of the latter, but rather as a specific geosystem, despite its much smaller spatial dimension.

Another significant physiognomic change in the landscape occurs to the east of the Monjolos karst, where the Serra do Espinhaço Meridional geosystem is located. This has the largest spatial dimension of the study area and presents some variations in the spatial distribution of ecological potential, on quartzite, and biological exploitation, native grassland. This geosystem is subdivided into the geofacies West Face, with stretches forming a morphodynamic mosaic tending towards instability; Interfluve, where pedogenesis predominates, and East Face, with a predominantly intergrade morphodynamic mosaic tending towards instability. Occurrences of other metasedimentary rocks embedded in the quartzites, especially on the west face, result in morphodynamic instability and moderate soil vulnerability.

The Serra do Espinhaço Meridional has ecological potential based on quartzite, proving to be resistant and with Natural Vulnerability classified as low, varying to very low or medium. However, biological exploitation, mostly composed of native grassland vegetation cover, means that the West Face and the Interfluve have high Environmental Vulnerability, while localised instances of anthropogenic action in the middle of the grassland cover mean that the East Face has very high Environmental Vulnerability. The ecological potential favours human activity aimed at tourism and inhibits activities for agricultural purposes.

Like the West Face, the Serra do Cabral geosystem has predominantly low Natural Vulnerability, with localised stretches in the other classes. Pedogenesis prevails over most of the terrain, where there are intergrade and moderately stable morphodynamic environments. However, there are occurrences of unstable media at moderate and medium levels in stretches of high and very high topographic roughness, where morphogenesis prevails. Despite the occurrence of young soils (Litholic or Quartzarenic Neosols), the soil is modelled in the moderately stable / vulnerable class.

The very location surrounded by the Rio das Velhas Plain is indicative of the environmentally highly resistant capacity. Despite similarities to the ecological potential of the Serra do Espinhaço Meridional, the location implies more resistant biological exploitation, made up of greater occurrences of savannah vegetation cover modelled with low or medium Environmental Vulnerability. Likewise, in the Rio das Velhas Plain, stretches with anthropogenic action consisting of planted eucalyptus forests were modelled with very low or low Environmental Vulnerability.

CHAPTER 5

FINAL CONSIDERATIONS

"In real life, things end with less format, they don't end at all. It's better that way. Striving for the exact is a mistake against us. Don't want to. Living is very dangerous... "

(J. Guimarães Rosa, O Grande Sertão: Veredas.)

The geoecological modelling of the landscape showed different results, although they were relatively compatible for the Corinto and Diamantina regions. The geosystem approach (BERTRAND, 2004) ensured compatibility by characterising the environmental aspects and analysing the elements of ecological potential, ecological exploitation and anthropic action, using transect sampling. Four geosystems were identified: Planície do Rio das Velhas, Carste de Monjolos, Serra do Cabral and Serra do Espinhaço Meridional, the latter comprising three geofacies (West Face, Interfluve and East Face).

It should be noted that the database, made available digitally by various specialised public institutions, required effort to consolidate, correct and validate before being submitted to the modelling applied. Even so, when compared with information produced by specific surveys, there were differences, partly due to the use of different spatial scales. The mapping of the soil cover of Minas Gerais, carried out by the DPS / UFV (2006) compares favourably with the field surveys undertaken in this research. However, this was not the case with the mapping of the vegetation cover of Minas Gerais (IEF, 2009) compared to the work of Rodrigues (2011) and Rodrigues and Travassos (2013), who identified Dry Forests over the carbonates in the Monjolos region.

In addition to these caveats regarding the consolidated digital data, a few observations can be made about the modelling based on Tricart (1977), Crepani et al. (2001) and Jansen (2013), as well as the potential uses of the study area.

- Modelling based on Tricart (1977) and Crepani et al. (2001):

The simple map algebras, composed of the addition of 03 variables, in the case of the morphodynamic modelling based on Tricart (1977), and 06 variables, in the case of the modelling proposed by Crepani et al. (2001) for natural vulnerability to soil loss, resulted in the landscape of the study area being classified in the respective intermediate classes. However, the morphodynamic modelling, which has 07 classes, showed a more precise intermediate class than the soil vulnerability modelling. The modelling proposed by Crepani et al. (2001), whose final model has 21 classes aggregated into 5 degrees of vulnerability, brings together more classes in the intermediate degree, resulting in greater generalisation.

The modelling by Crepani et al. (2001) provided support for the other modelling, based on Tricart (1977) and proposed by Jansen (2013), especially in the parameterisation of the lithological aspect using the variable Degree of Rock Cohesion (GCR). However, this parameterisation, which is assumed to have a subjective content, covers a relatively small set of rock types, leaving gaps that end up being filled by other researchers (e.g.: RIBEIRO; CAMPOS, 2007; SANTOS; SOBREIRA, 2008; MESQUITA; ASSIS; SOUZA, 2010; JANSEN, 2013), also with a subjective content.

The GCR variable is not the only one or necessarily the most suitable for representing the lithological aspect in a map algebra, as it refers to mechanical conditions and does not take into account the geochemical conditions of the rocks. Including "faults and fractures" as a variable would improve the modelling by considering fractures and faults as inducing greater geological fragility in their surrounding areas. Thus, the same fractured lithological body would have different levels of fragility, more accentuated around faults and fractures or less accentuated in stretches away from these structural elements. Finally, the inclusion of "lithological structure" would also improve the modelling by considering the different types of structure as parameters of greater or lesser environmental fragility.

- Modelling by Jansen (2013):

Among the modelling techniques applied, the one proposed by Jansen (2013) best suited the geosystem approach. Its map algebra is weighted, made up of the addition of five variables with different weights, and resulted in the landscape in the study area being better distributed in the Natural Vulnerability 05 classes, which corresponds to measuring ecological potential. With the simple addition of a sixth variable (vegetation cover), the Environmental Vulnerability results, corresponding to the combination of ecological potential with biological exploitation and anthropogenic action, were concentrated in the very high class, the most fragile possible.

Given the lithological diversity, typical of the contact between metasedimentary rocks of the Espinhaço Supergroup and sedimentary rocks of the Bambuí Group, the Monjolos karst geosystem (on marly limestones of the Bambuí Group) stands out for its greater environmental fragility. It has the most unstable morphodynamic conditions, the most prominent soil vulnerability and very high Natural and Environmental Vulnerabilities. In the three models generated, the differences between the Monjolos Karst and the Rio das Velhas Plain allow the two to be recognised as different geosystems. The Monjolos Karst is not a geofacies of the Rio das Velhas Plain.

The Serra do Espinhaço Meridional has different dynamics on its western and eastern slopes, which gradually advance over the interfluve endowed with a certain natural stability. In this geosystem, on resistant quartzites of the Espinhaço Supergroup, more stable morphodynamic conditions predominate, or on the threshold of stability, although unstable conditions occur on the West Face. This geofacies showed the most eclectic results of the study area in the three modelling approaches applied, standing out for its environmental complexity. The Serra do Cabral geosystem showed similar results to the West Face.

The Interfluve geofacies, contrary to what the high altimetric levels might suggest, showed moderate stable morphodynamics and at the threshold of vulnerability to soil loss, with predominantly low Natural Vulnerability, varying to very low or medium. However, the vegetation cover means that Environmental Vulnerability is predominantly high with very high enclaves. Among the geofacies of the Serra do Espinhaço Meridional, the East Face is the most environmentally fragile, where very high Environmental Vulnerability predominates because Natural Vulnerability is mostly medium.

The use of the variable "Topographic Roughness Concentration Index" by Jansen (2013), which corresponds to a synthesis of the three variables used by Crepani et al. (2001), facilitates the handling and application of the relief aspect in geoecological modelling.

- Proposals for new Natural Vulnerability modelling:

Respecting the "geology" variable as having the greatest weight in Natural Vulnerability modelling (JANSEN, 2013), it is pertinent to check possible calibrations of this model through different weights for the other variables (relief, soil, climate and speleological potential). Figure 38 shows variations in the Natural Vulnerability model in the study area, resulting from equal weights between the "pedology" and "Topographic Roughness Concentration Index - RCI" variables in the map algebra:

$$VN = GEO*0.35 + PED*0.20 + ICR*0.20 + IP*0.15 + PCAV*0.10$$

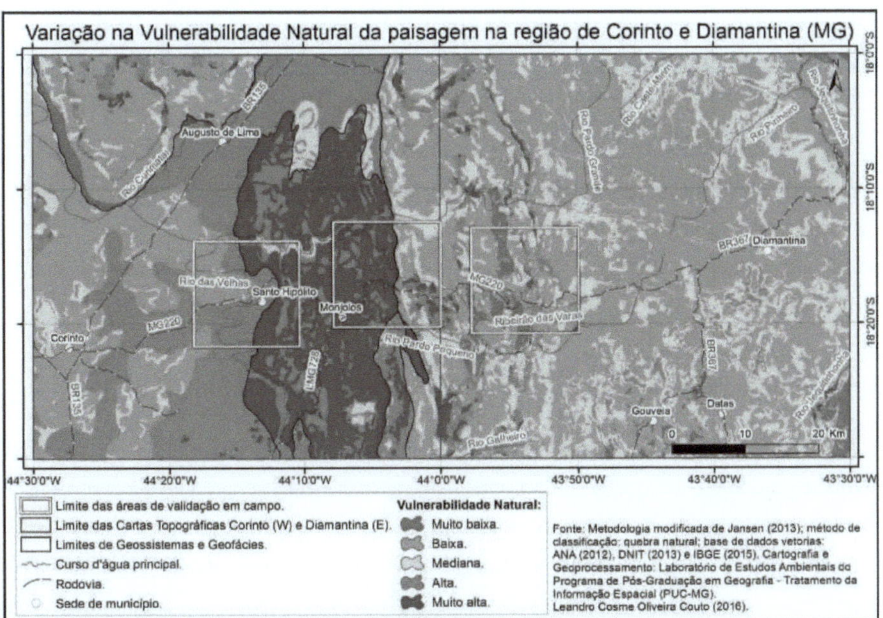

Figure 38 - Map of variation in the Natural Vulnerability model in the study area.
Source: Prepared by the author.

Compared to the results of the original modelling proposed by Jansen (2013) (Figure 34), the results of this variation in the calibration of the weights differ significantly due to the general increase in the natural vulnerability of the study area. There is a decrease in stretches classified as having very low Natural Vulnerability, especially in the three geofacies of the Serra do Espinhaço Meridional geosystem, which now have stretches predominantly classified as having low Vulnerability. Accompanying this general increase, the effects of this variation in calibration increase the Natural Vulnerability of the Rio das Velhas Plain and Monjolos Karst geosystems, which are almost entirely classified as having very high Natural Vulnerability. The results corroborate the main finding of the original modelling that the Monjolos karst is the most fragile geosystem in the study area.

Another Natural Vulnerability modelling proposal that could be studied and researched involves more significant modifications to the equation, changing the weights and repositioning and/or including other variables. The repositioning and weighting of the variables takes into account the geological and climatic conditions of the relief and pedological cover:

$$VN = GEO*0.35 + IP*0.25 + PED*0.20 + ICR*0.20$$

Where:

VN = Natural Vulnerability;
GEO = Geology, arithmetic mean of Degree of Rock Cohesion (GCR), Faulting and

Fracturing (FF), Lithological Structure (EL) and Potential for Cavity Occurrence (PCAV);
IP = Rainfall Intensity, the arithmetic mean between the Annual Rainfall Quantity (QAP) and the Annual Rainfall Distribution (DAP);
PED = Pedology, represented by Soil Maturity;
ICR = Topographic Roughness Concentration Index.

In this proposal, the parameterisations of the variables already used in other modelling should be improved, and the parameterisations of the new variables should be consolidated. Along with the refinements and improvements that could be made to the modelling applied, the assessment of potential uses in the study area points to the need for further studies and research.

- Potential uses in the study area:

The results of the geoecological modelling applied to the landscape of the Corinto and Diamantina regions allow for assessments of potential uses in the study area. To do this, however, it is necessary to revise the digital databases, taking into account the original scales used. There is also a need for greater refinement in the set of data relating to anthropogenic action in each mapped geosystem.

As we found out in the field, in the Rio das Velhas Plain and the Monjolos karst there are farming activities, while in the Serra do Espinhaço Meridional there are tourism activities due to the abiotic and biotic aspects, which are more prominent, as well as the historical and cultural aspects.

Potential uses in each geosystem can be assessed individually. However, an integrated assessment is more appropriate. Thus, the agricultural potential of the Rio das Velhas Plain and the Monjolos Karst should be combined with the geoecological conditions of morphodynamic instability and vulnerability to natural and environmental soil loss. In turn, the tourism potential of the Serra do Espinhaço Meridional can be accentuated by integrating it with the geotourism potential of the Monjolos karst, which is still little explored.

REFERENCES

AB'SABER, A. N. Os domínios de natureza no Brasil: potencialidades paisagísticas. 2. ed. São Paulo: Nacional, 2003.

ABREU, A. The Espinhaço Supergroup of the Serra do Espinhaço Meridional (Minas Gerais): the rift, the basin and the orogen. Geonomos. 3 (1): 01-18, 1995.

AMORIM FILHO, O. B. The plurality of Geography and the need for cultural approaches. In: *Caderno de Geografia, v.* 16, n° 26. Belo Horizonte, 2006. p. 35-56.

ANA, National Water Agency. *Shapefile of the national hydrographic network.* 2012. Original scale: 1:1.000.000. Available at http://metadados.ana.gov.br/geonetwork/srv/pt/main.home. Accessed on: January 2016.

ANDREYCHOUK, V. et al. *Karst in the Earth 's Crust: its distribution and principal types*. Poland: University of Silesia / Ukrainian Academy of Sciences / Tavrichesky National University-Ukrainian Institute of Speleology and Karstology, 2009.

BARROSO, L. C. and ABREU, J. F. (Orgs) *Geography, spatial analysis models and GIS*. Belo Horizonte: PUC Minas, 2003.

BECKER, B. K. Geopolitics at the turn of the millennium: logistics and sustainable development. In: CASTRO, I. E. de; GOMES, P. C. da C; Corrêa, R. L. *Geografia: conceitos e temas*. 15ª ed. Rio de Janeiro: Bertrand Brasil, 2012.

BERTRAND, G. Landscape and global physical geography: a methodological outline. Translation: Olga Cruz. *RA' E GA magazine*. UFPR Publishing House. Curitiba, n. 8, p. 141-152. 2004.

CECAV, National Centre for Cave Research and Conservation. *Shapefile of the distribution of natural underground cavities in Brazil*. 2016. Original scale: 1:2,500,000. Available at: http://www.icmbio.gov.br/CECAV/downloads/mapas.html. Accessed on: January 2016.

CHRISTOFOLETTI, A. *Systems analysis in Geography*. São Paulo: HUCITEC, USP, 1979.

CHRISTOFOLETTI, A. *Modelling environmental systems*. São Paulo: Edgard Blucher, 1999.

CODEMIG, Minas Gerais Economic Development Company. *Shapefile of the lithology of the Topographic Map of Diamantina (MG)*. Original scale: 1,100,000. Available at: http://www.portalgeologia.com.br/index.php/mapa/#downloads-tab. Accessed on: January 2016.

CONAMA, National Environment Council. Resolution 357 of 17 March 2005. Amended by Resolution 410/2009 and 430/2011. Available at: http://www.mma.gov.br/port/conama/res/res05/res35705.pdf. Accessed on: 28/03/2016.

COPAM, Minas Gerais State Environmental Policy Council. Normative Deliberation No. 20, of 24 *June 1997*. Provides for the classification of waters in the Velhas river basin. Available at: http://www.siam.mg.gov.br/sla/download.pdf?idNorma=115. Accessed on: 28/03/2016.

CPRM, Companhia de Pesquisa de Recursos Minerais / Serviço Geológico do Brasil. *Shapefile of the lithology of the Topographic Map of Corinto (MG)*. Original scale: 1,100,000. Available   at: http://geobank.cprm.gov.br/pls/publico/geobank.download.downloadVetoriais7p   webmap=N   &p usuario=1. Accessed on: January 2016.

CREPANI, E. et al. *Remote sensing and geoprocessing applied to ecological-economic zoning and land-use planning*. São José dos Campos: INPE, 2001.

DNIT, National Department of Infrastructure and Transport. *Shapefile of national motorways*. 2013. Original scale: 1:1,000,000. Available at http://www.dnit.gov.br/mapas- multimodal/shapefiles. Accessed on: October 2015.

Morphoclimatic and phytogeographic domains of Brazil. Professor Murilo Guirro's blog. [S.l.]: Prof Murilo Guirro, 2016. Available at: http://professormuriloguirro.blogspot.com.br/p/aulas-minhas.html/. Accessed on: 05 April 2016.

DPS / UFV, Department of Soils, Federal University of Viçosa. *Shapefile of the distribution of soils in Minas Gerais*. 2006. Original scale: 1:650.000. Available at: http://www.dps.ufv.br/7page id=742. Accessed in January 2016.

DSG, Geographical Service Department of the Brazilian Army. *Corinto Topographic Map,* SE-23-Z-A-II (MI 2422). Rio de Janeiro: IBGE, 1977. Scale 1:100,000.

ESRI, *Environmental Systems Research Institute.* ArcGis 10.2. Commercial software. Available from the Environmental Studies Laboratory of the Postgraduate Programme in Geography - Treatment of Spatial Information (PUC-MG). Accessed in: 2015 and 2016.

FABRI, F. P. *Study of the quartzite caves of the Itambé do Mato Dentro region, Serra do Espinhaço Meridional (MG).* Master's dissertation. UFMG: Belo Horizonte, 2011.

FIBGE, Brazilian Institute of Geography and Statistics Foundation. *President's Resolution No. 01/2005* (R. Pr. - 1/2005). Changes the characterisation of the Brazilian Geodetic System. Date: 25/02/2005. Available at: ftp://geoftp.ibge.gov.br/documentos/geodesia/projeto_geodesic-reference-change/legislation/rpr_01_25feb2005.pdf. Accessed on: 13 October 2005.

FOGAÇA, A. C. C. *Diamantina.* Geological Map. Scale 1:100,000. Belo Horizonte: Minas Gerais Economic Development Company (CODEMIG), 2011.

GOMES, M. *Methodology for identifying vulnerable areas for the conservation of the Brazilian Speleological Heritage.* Monograph (Specialisation in geoprocessing) - UFMG, Belo Horizonte, 2010.

GONTIJO, B. M. A geography for the Espinhaço Chain. Megadiversity. v. 4, n. 1-2, p. 7-15, 2008.

GOOGLE INC. Google Earth. Free software. Available for download at: https://www.google.com.br/earth/download/ge/agree.html. Accessed on 04/04/2016.

GUIMARAES, R. L. Geomorphological mapping of the karst of the Monjolos region, Minas Gerais Master's thesis. PUC Minas: Belo Horizonte, 2012.

GUIMARAES, R. L., TRAVASSOS, L. E. P. e LINKE, V. A Geografia Cultural do carste tradicional carbonático de Monjolos, MG: uma primeira aproximação. In: *Proceedings of the 31st Brazilian Congress of Speleology.* Brazilian Speleological Society: Ponta Grossa (PR), 21-24 July 2011.

HAGGETT, P. and CHORLEY, R. J. Models, paradigms and the New Geography. In. CHORLEY, R. J. and HAGGETT, P. *Physical and information models in Geography.* Rio de Janeiro: Livros Técnicos e Científicos, 1-19, 1975.

IBGE, Brazilian Institute of Geography and Statistics. *Diamantina Topographic Map,* SE-23- Z-A-III (MI 2423). Rio de Janeiro: IBGE, 1977. Scale 1:100,000.

IBGE, Brazilian Institute of Geography and Statistics. *Brazil Climate Map.* Rio de Janeiro: IBGE, 1978. Scale 1: 5,000,000.

IBGE, Brazilian Institute of Geography and Statistics. *Map of relief units of Brazil.* 2. ed. Rio de Janeiro: IBGE, 2006. Scale 1: 5,000,000.

IBGE, Brazilian Institute of Geography and Statistics. *Technical manual of pedology.* 2ª ed. Rio de Janeiro: IBGE, 2007.

IGA, Institute of Applied Geosciences. *Digital Atlas of Minas Gerais.* Geomorphology. Available at: http://www.iga.mg.gov.br/mapserv_iga/atlas/TutorialPDF/7-

Geomorphology.pdf. Accessed on 09/03/2016.

JANSEN, D. C. *Brazilian map of potential cave occurrences.* In: IX ANPEGE National Meeting, 2011, Goiânia. IX ENANPEGE - National Meeting of the National Association for Postgraduate Studies and Research in Geography, 2011.

JANSEN, D. C. *Environmental analysis of the Morro da Pedreira environmental protection area and the Serra do Cipó National Park for the protection of speleological heritage.* Master's dissertation. PUC Minas: Belo Horizonte, 2013.

KNAUER, L. G.; et al. *Corinto.* Geological Map. Scale 1:100,000. Belo Horizonte: Companhia de Pesquisa de Recursos Minerais / Serviço Geológico do Brasil (CPRM), 2011.

MENESES, I. C. R. R. C. *Geosystemic analysis in the Carste environmental protection area (APA) of Lagoa Santa, MG.* Master's thesis - PUC Minas, 2003.

MESQUITA, C.; ASSIS, A. Q. S. de; SOUZA, R. M. de. Natural vulnerability to soil loss in the Sagrado River basin - Morretes/PR. In: *Revista de Geografia. Recife:* UFPE - DCG/NAPA, special v. VIII SINAGEO, n. 2, Sep. 2010.

MONTEIRO, C. A. de F. *Geosistemas: a história* de *uma procura.* 2 ed. São Paulo: Contexto, 2001.

PILO, L. B. Literature Review - Karst Geomorphology. In: *Revista Brasileira de Geomorfologia.* São Paulo: University of São Paulo, Volume I, No. 1 (2000): 88-102.

PISSINATI, M. C. and ARCHELA, R. S. Geosistema território e paisagem - Método de estudo da paisagem rural sob a ótica Bertrandiana. In: *Geography.* Universidade Estadual de Londrina (UFL) - Departamento de Geociências, v. 18, n. 1, jan./jun. 2009.

REIS, R. J.; GUIMARÃES, D. P.; LANDAU, E. C. Rainfall in Minas Gerais. Belo Horizonte: Ed. PUC Minas, 2012.

ROCHA, L. C. *Geochemical denudation in the evolution of the Serra do Espinhaço Meridional, MG - Brazil.* PhD Thesis. UFMG: Belo Horizonte, 2011.

RODRIGUES, B. D. *Identification and mapping of dry forests associated with the carbonate karst of Santo Hipólito and Monjolos, Minas Gerais* / Master's thesis. PUC Minas: Belo Horizonte, 2011.

RODRIGUES, B. D. and TRAVASSOS, L. E. P. Identification and mapping of dry forests associated with the carbonate karst of Santo Hipólito and Monjolos. *Mercator - Revista de Geografia da UFC (Federal University of Ceará).* Fortaleza, v. 12, n. 29, p. 233-256, Sep./Dec. 2013.

SAADI, A. The geomorphology of the Serra do Espinhaço in Minas Gerais and its margins. In: *Geonomos* 3 (1): 41-63, 1995.

SANTANNA NETO, J. L. Decalogue of the climatology of southeastern Brazil. In: *Revista Brasileira de Climatologia,* v. 1, n° 1: 43-60, 2005.

SANTOS, C. A. dos; SOBREIRA, F. G. Analysing the fragility and natural vulnerability of land to erosion processes as a basis for land-use planning: the case of the Córrego Carioca, Córrego do Bação and Ribeirão Carioca basins in the Upper Rio das Velhas region - MG. In: *Revista Brasileira de Geomorfologia,* v.9, n.1, p.65-73, 2008.

SOTCHAVA, V. B. The study of the geosystem. In: *Methods in Question (No. 16)*. Geographical Institute of the State of São Paulo, p. 1-49, 1977.

TRAVASSOS, L. E. P., GUIMARÃES, R. L. and VARELA, I. D. Karst areas, caves and the Royal Road. In: *Research into Tourism and Karst Landscapes*. Campinas: SeTur/SBE, 1(2), 2008.

TRAVASSOS, L. E P. *Characterisation of the karst of the Cordisburgo region*, Minas Gerais, Brazil. Belo Horizonte: Tradição Planalto, 2010.

TRICART, J.   Ecodynamics.   Rio de Janeiro: IBGE, 1977. Available at: http://biblioteca.ibge.gov.br/visualizacao/monografias/GEBIS%20-%20RJ/ecodinamica.pdf. Accessed on: 13 November 2015.

TROMPETTE, R. R., et al. The Brazilian São Francisco Craton - A review. In: Revista Brasileira de Geociências. São Paulo, 22(4):481-486, December 1992.

TROPPMAIR, H. Landscape ecology: a retrospective. Proceedings of the 1st Forum for Debates on Landscape Ecology and Environmental Planning. Rio Claro: Unesp, 2000a. Available at http://www.seb-ecologia.org.br/forum/art24.htm. Accessed on: 01 December 2015.

TROPPMAIR, H. Geosistemas e Geossistemas paulistas. Rio Claro, 2000b.

TROPPMAIR, H; GALINA, M. H. Geosystems. In: *Mercator - Journal of Geography of the UFC (Federal University of Ceará)*. Fortaleza, v.5, n. 10, p. 79-88, 2006.

UHLEIN, A., TROMPETTE, R. & EGYDIO-SILVA, M. Superposed rifting and inversion tectonics at the southeastern edge of the São Francisco Craton. In: *Geonomos* 3 (1): 99107, 1995.

ZEE-MG (a), Ecological-Economic Zoning of Minas Gerais. *Shapefile of the 2009 Mapping of Vegetation Cover*. Produced by the Minas Gerais State Forestry Institute (IEF), 2009. Original scale: 1.65.000. Available at: http://geosisemanet.meioambiente.mg.gov.br/zee/. Accessed on: January 2016.

ZEE-MG (b), Ecological-Economic Zoning of Minas Gerais. Shapefile of contour lines in the state of Minas Gerais. Produced by the Geographical Service Department of the Brazilian Army (DSG) and the Brazilian Institute of Geography and Statistics (IBGE). Original scales: 1:100.000   e 1:50.000. Available at: http://geosisemanet.meioambiente.mg.gov.br/zee/. Accessed on: January 2016.

APPENDICES

Figure 39 - Magnification: NNW view of Control Point 04 (18.28° S and 44.04° W).
Source: Prepared by the author, 19/06/2015.

Figure 40 - Magnification: ESE view of Control Point 05 (18.29° S and 43.84° W).
Source: Prepared by the author, 19/06/2015.

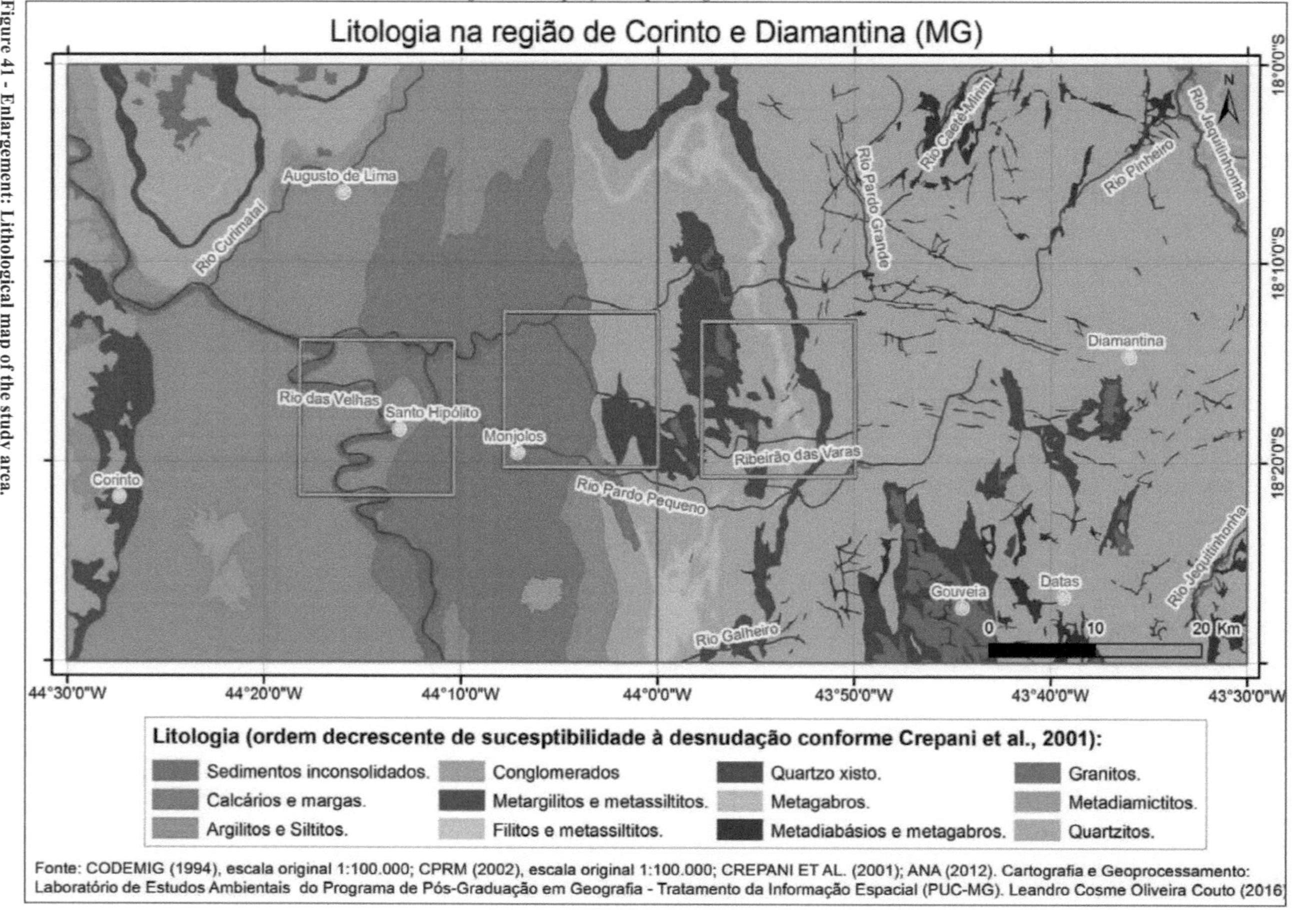

Figure 41 - Enlargement: Lithological map of the study area.
Source: Prepared by the author.

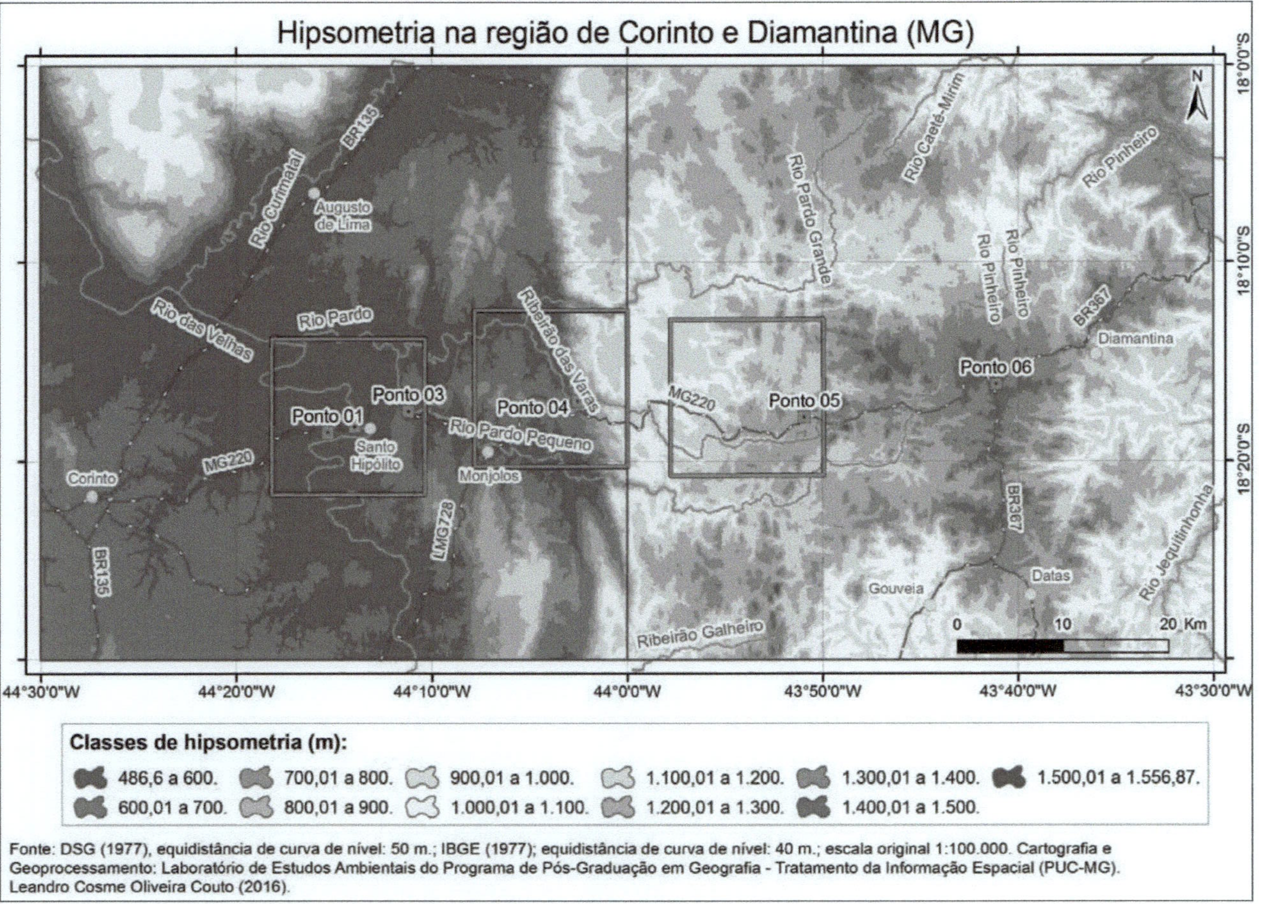

Figure 42 - Enlargement: Hypsometric map of the study area.

91

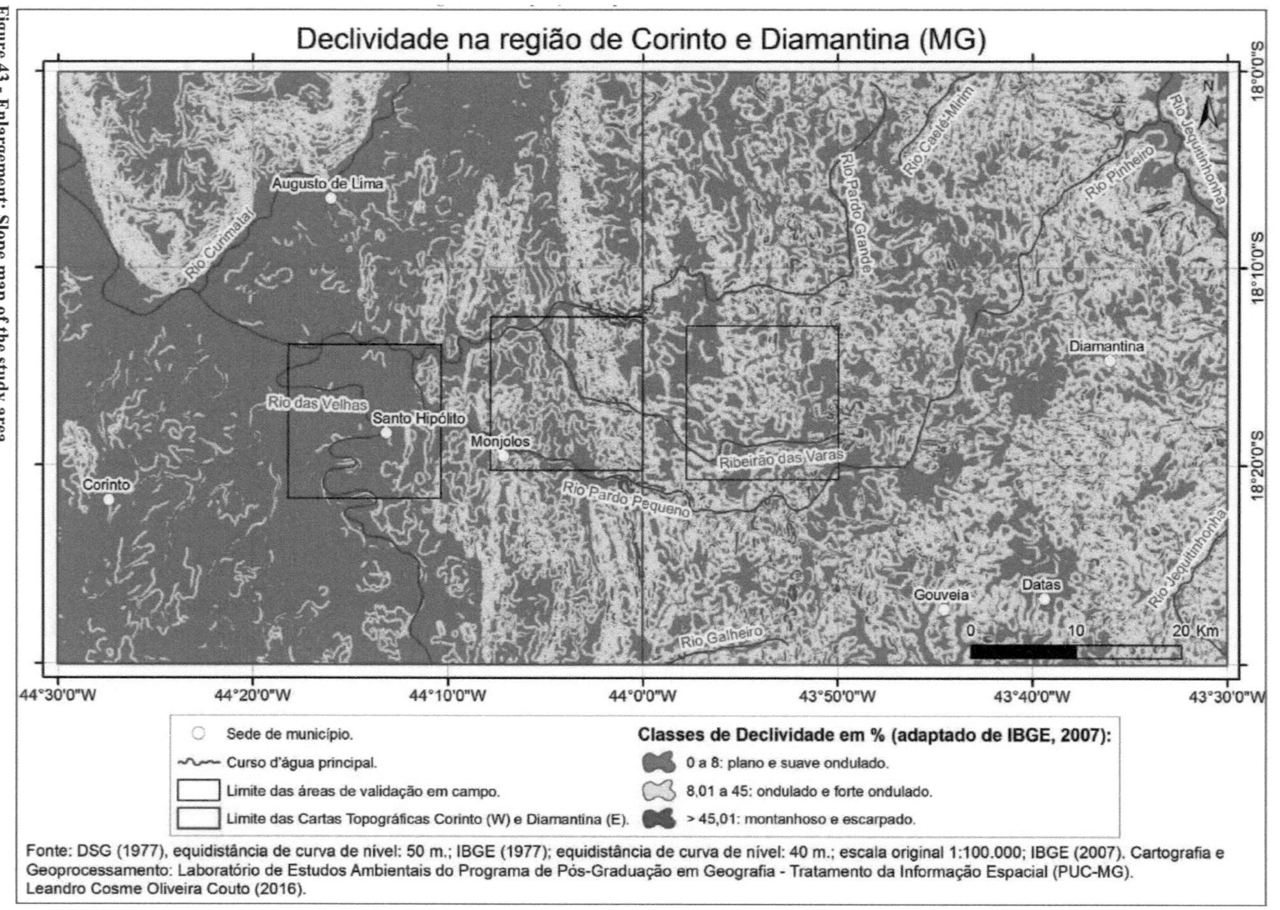

Figure 43 - Enlargement: Slope map of the study area.

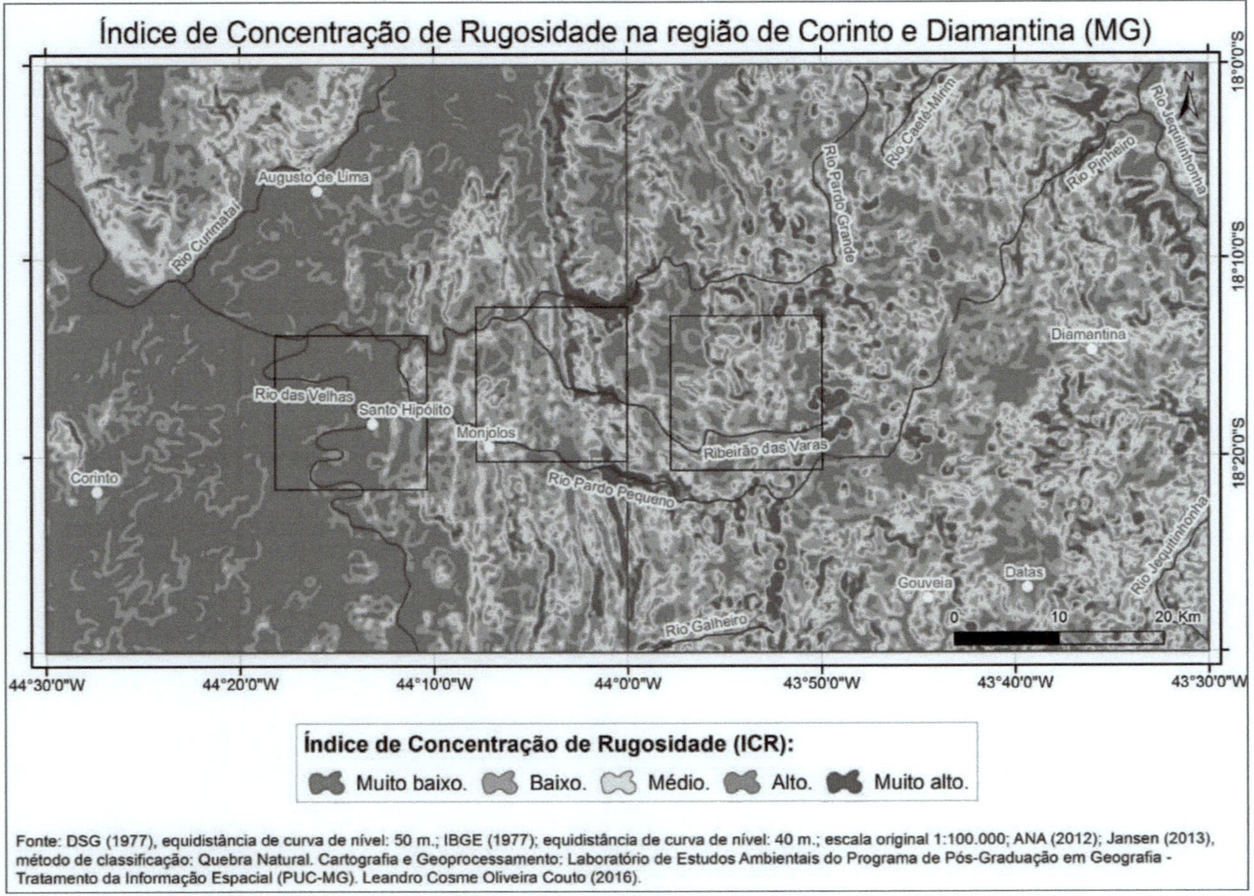

Fonte: DSG (1977), equidistância de curva de nível: 50 m.; IBGE (1977); equidistância de curva de nível: 40 m.; escala original 1:100.000; ANA (2012); Jansen (2013), método de classificação: Quebra Natural. Cartografia e Geoprocessamento: Laboratório de Estudos Ambientais do Programa de Pós-Graduação em Geografia - Tratamento da Informação Espacial (PUC-MG). Leandro Cosme Oliveira Couto (2016).

Figure 44 - Enlargement: Topographical roughness map of the study area.

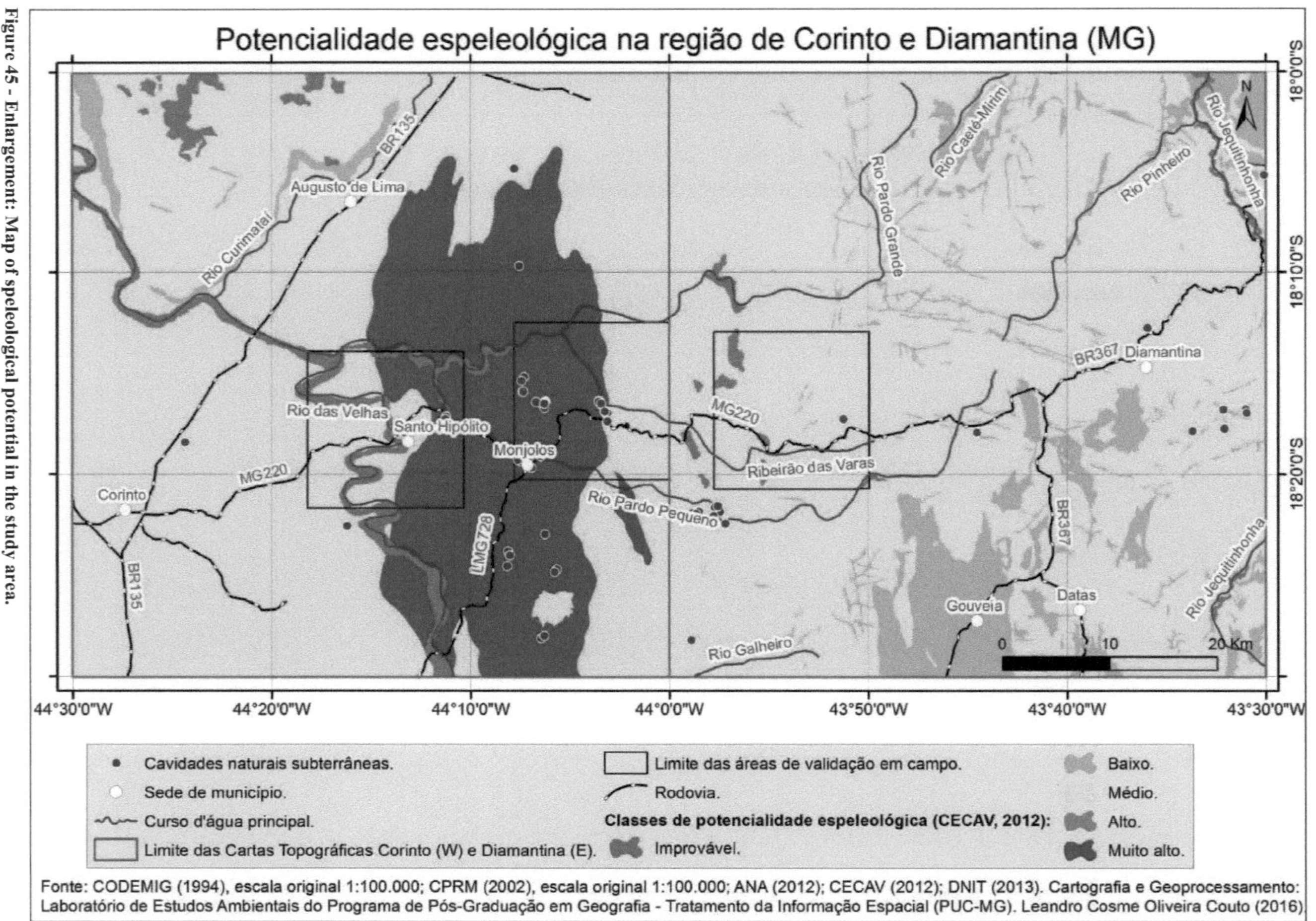

Figure 45 - Enlargement: Map of speleological potential in the study area.

94

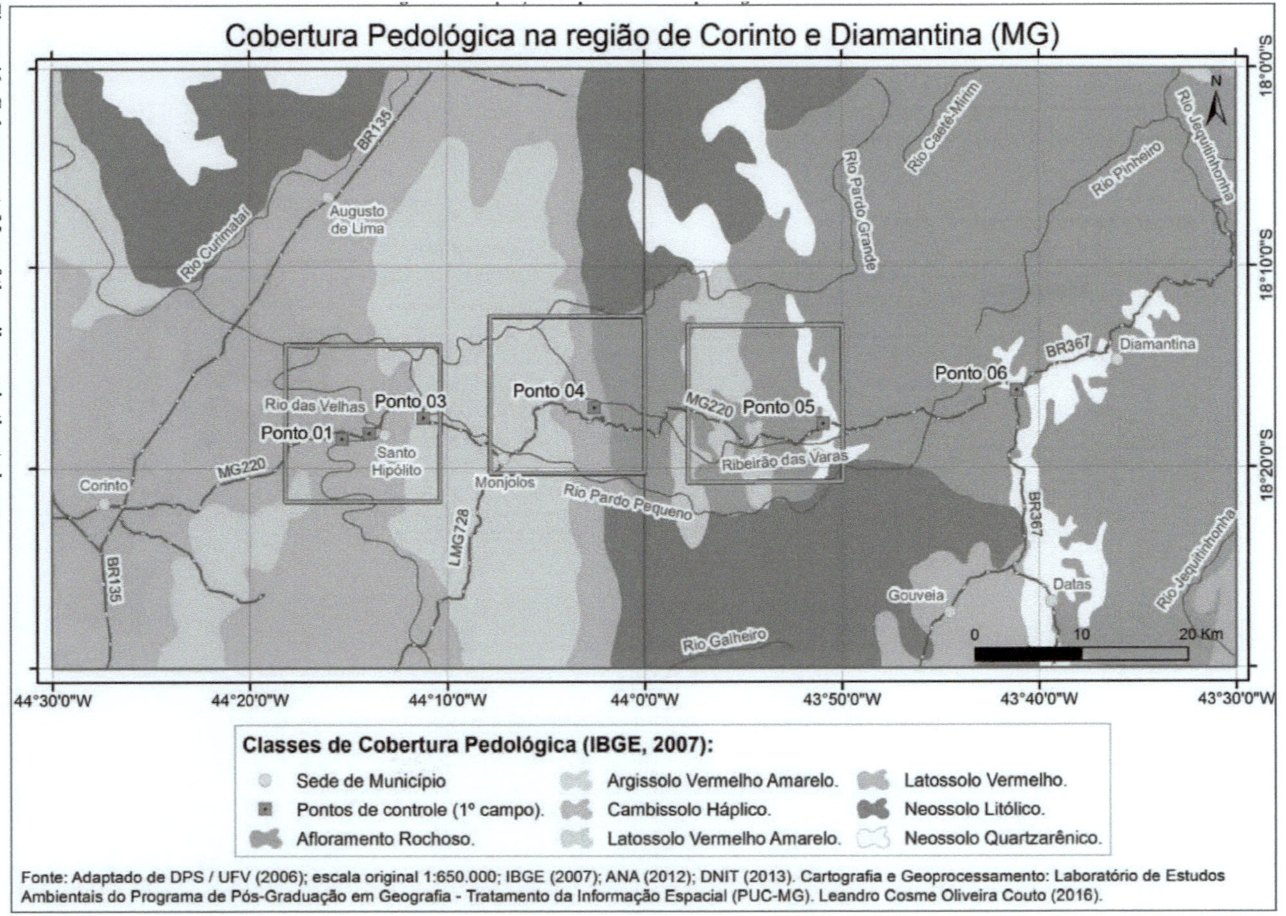

Figure 46 - Enlargement: Map of the soil cover in the study area.

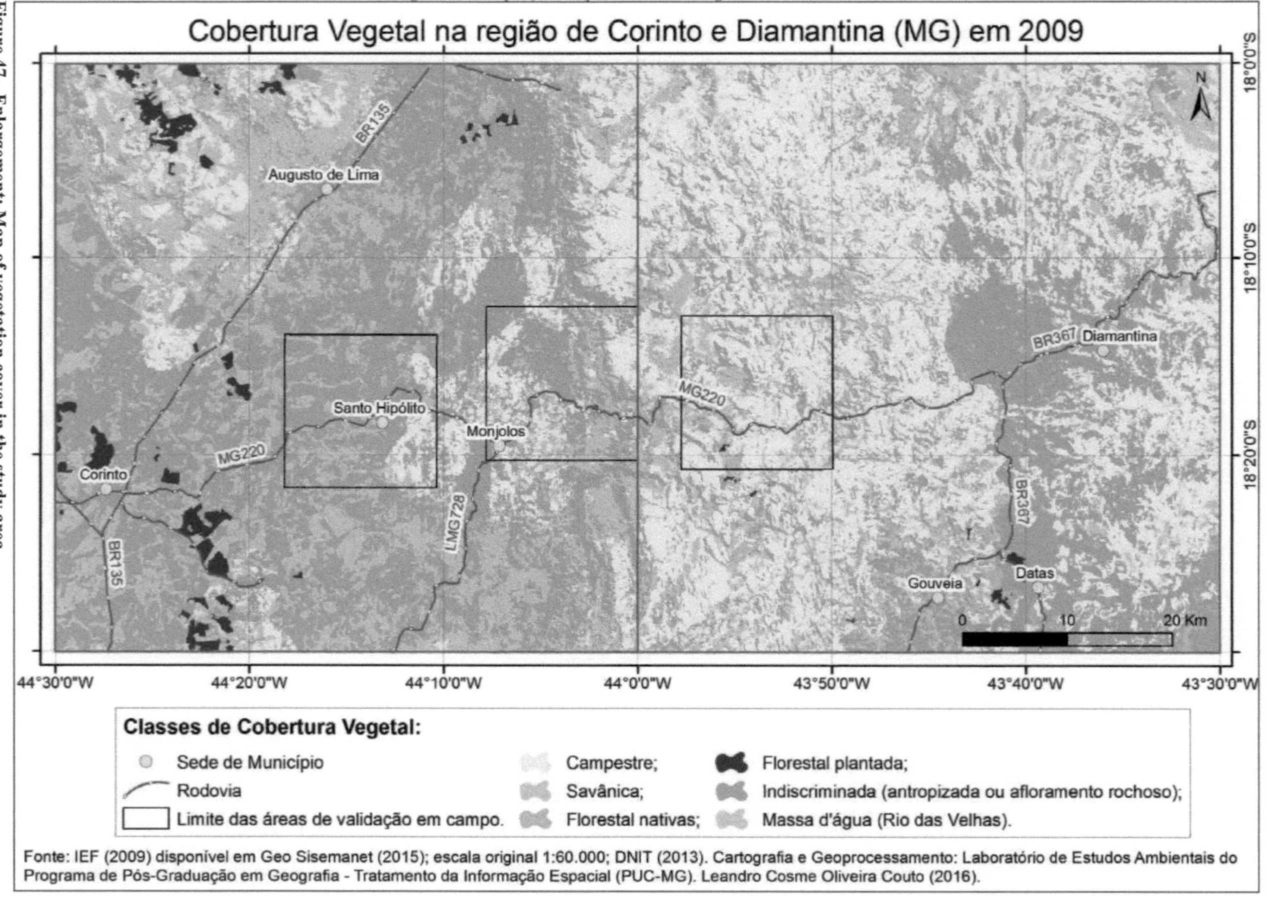

Figure 47 - Enlargement: Map of vegetation cover in the study area.

96

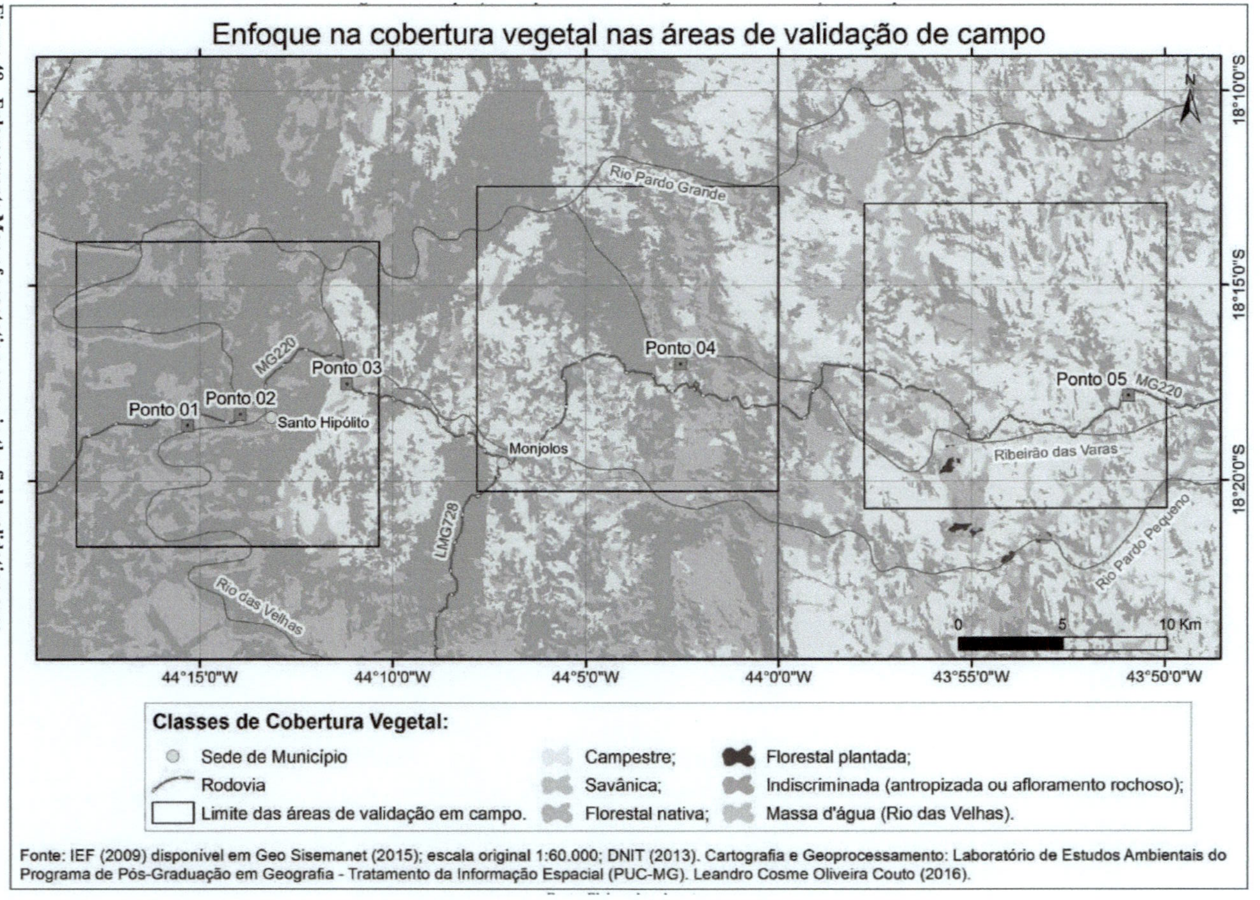

Figure 48 - Enlargement: Map of vegetation cover in the field validation area.

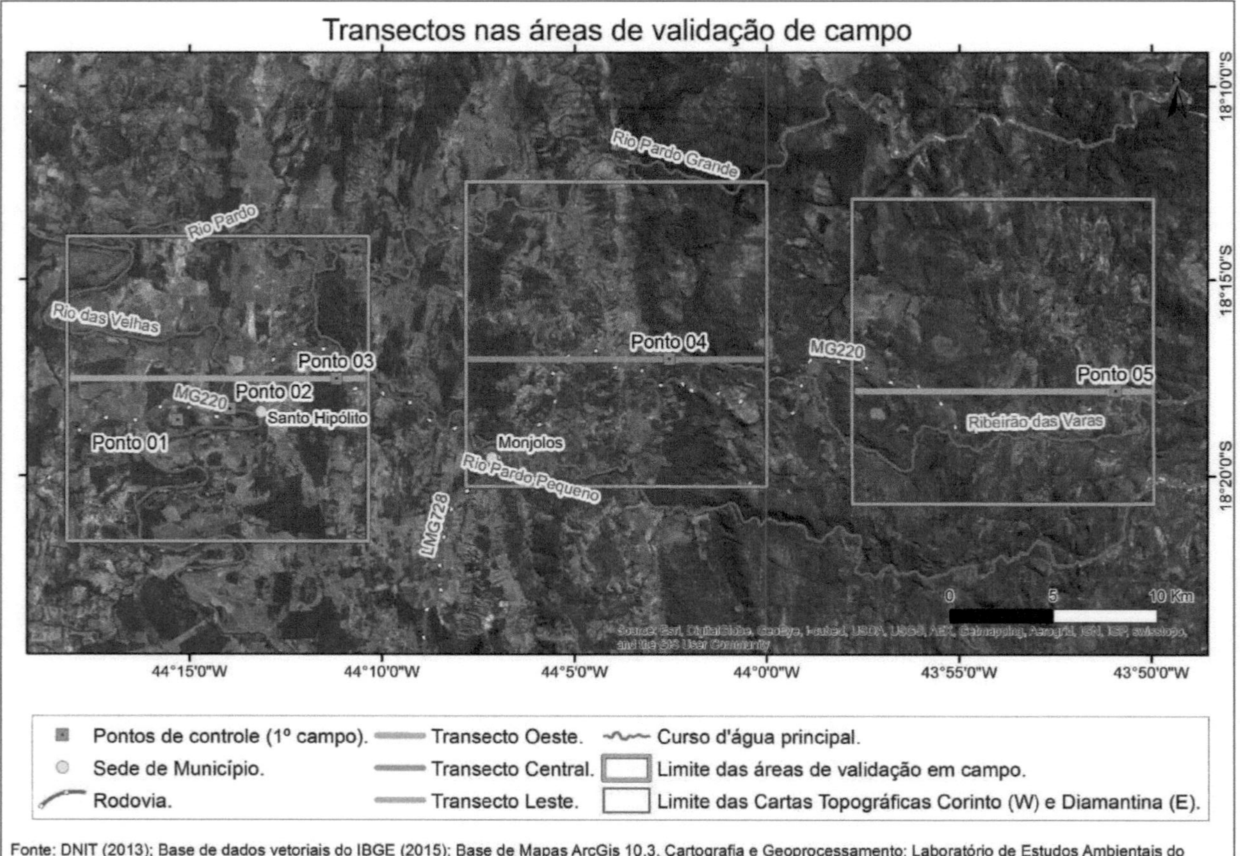

Figure 49 - Enlargement: Map of the transects in the field validation areas.

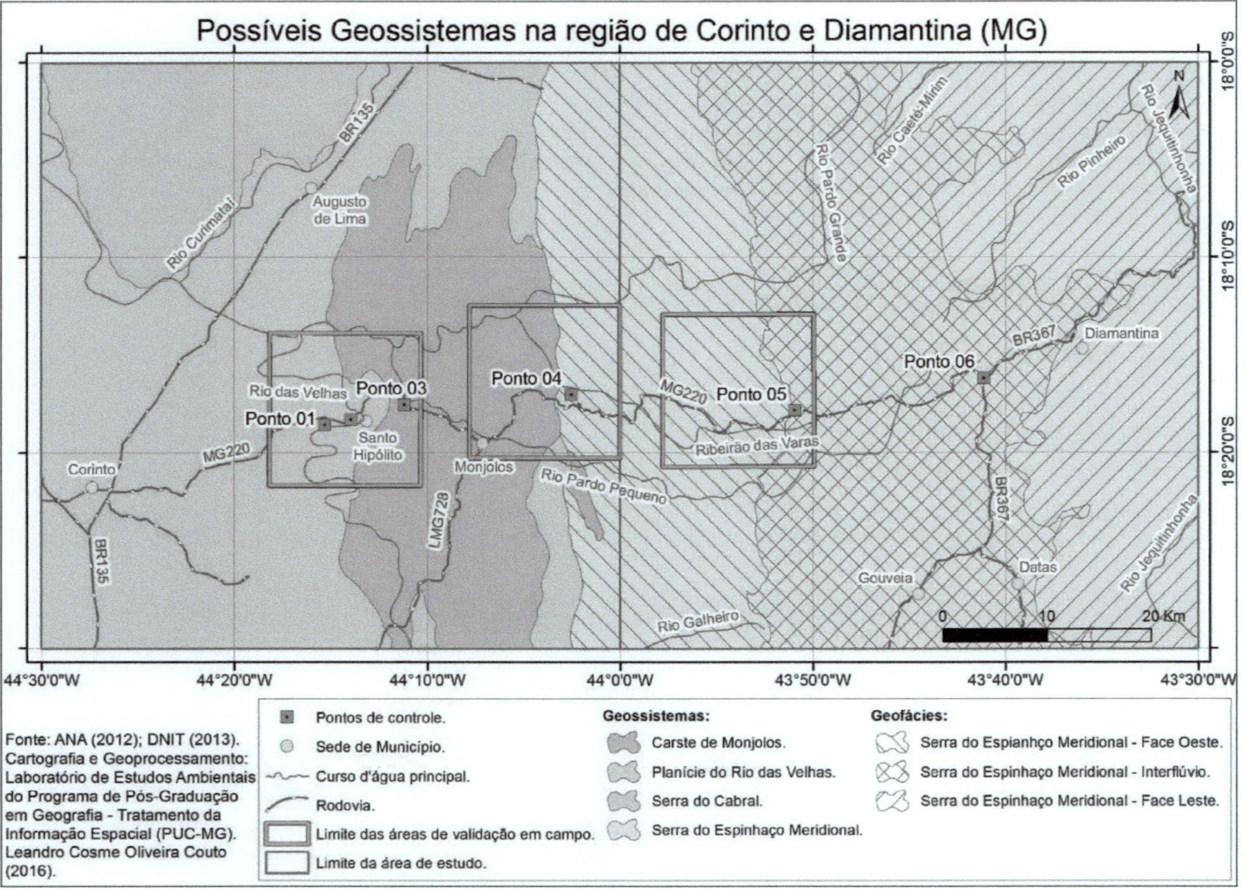

Figure 50 - Enlargement: Map of Geosystems in the study area.

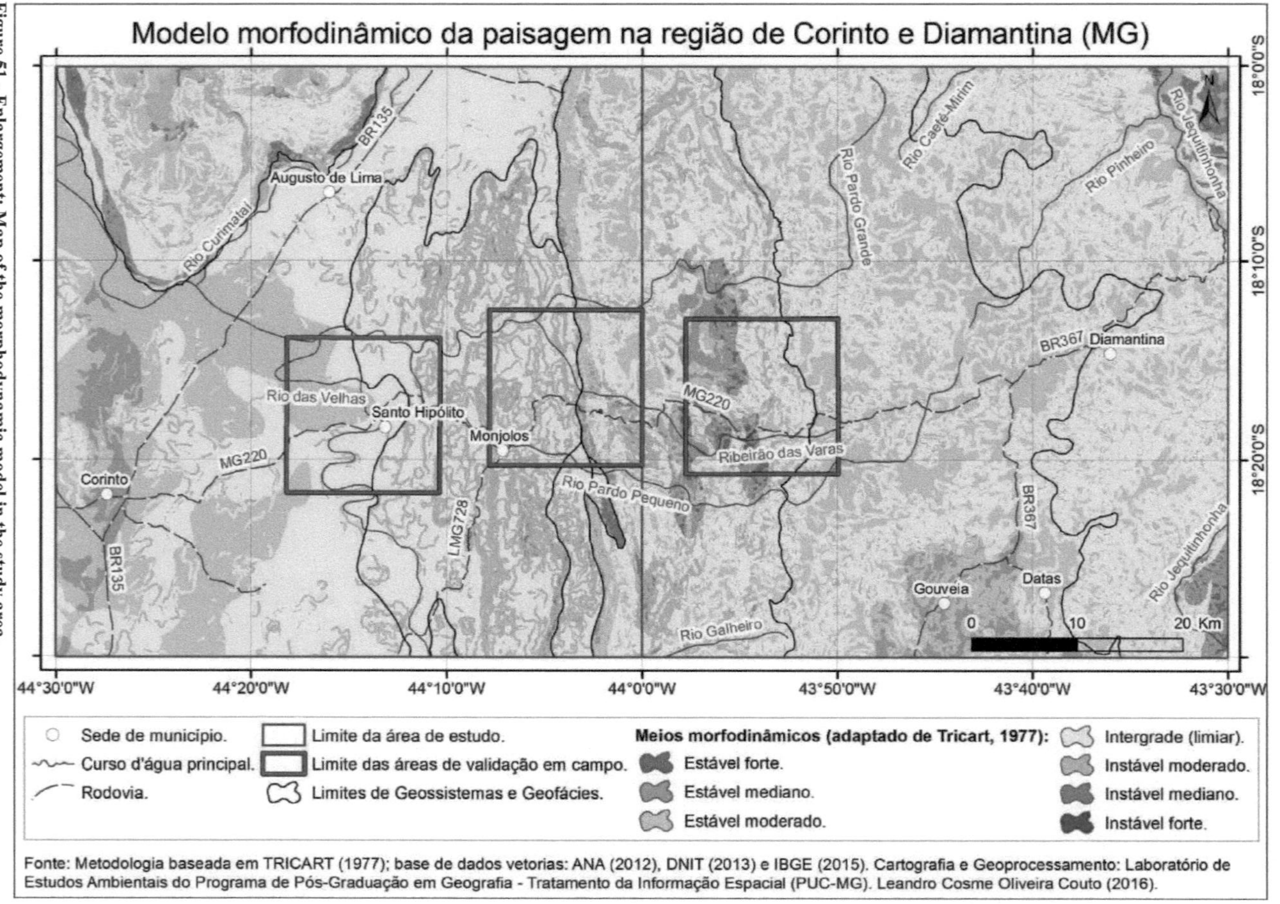

Figure 51 - Enlargement: Map of the morphodynamic model in the study area.

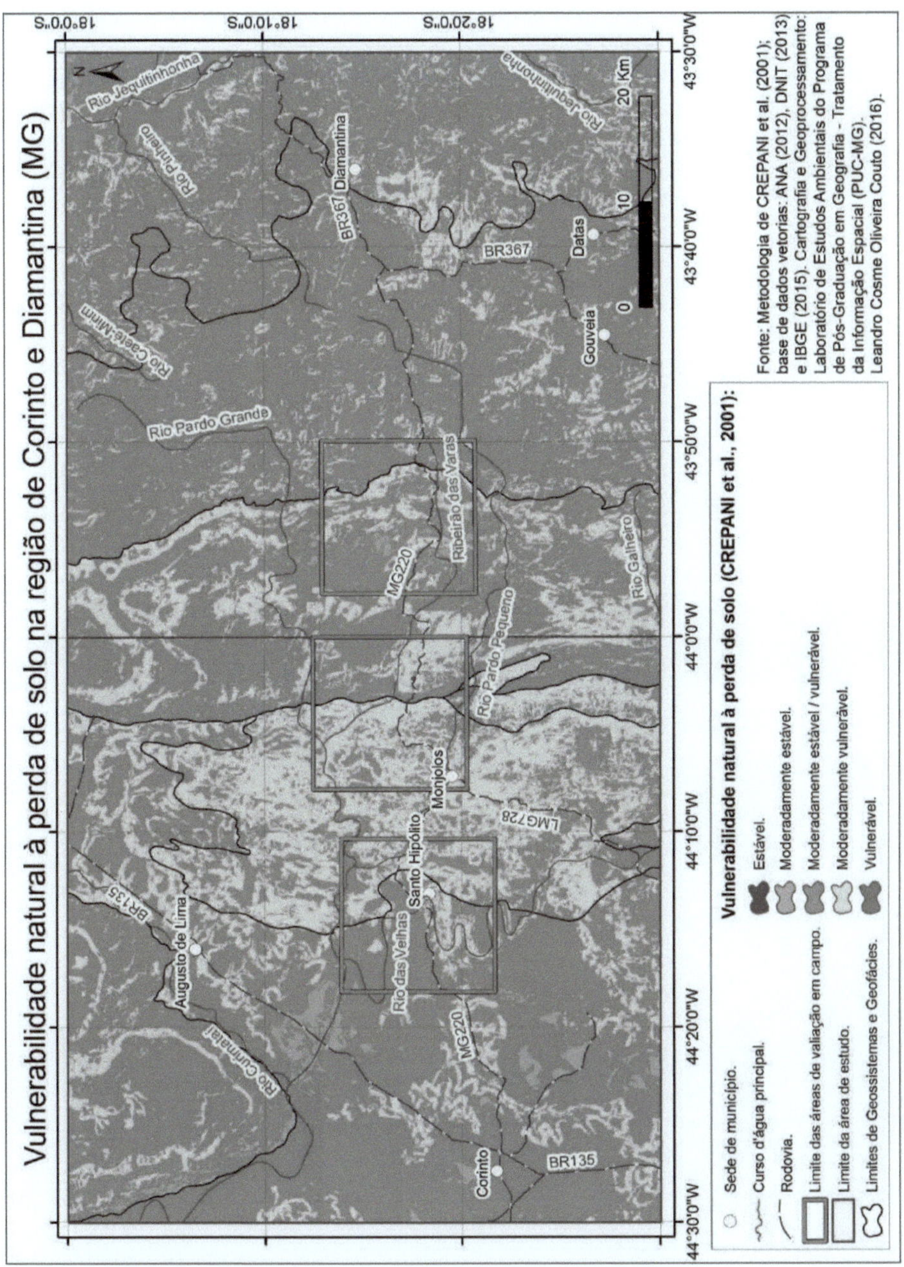

Figure 52 - Enlargement: Map of natural vulnerability to soil loss in the study area.

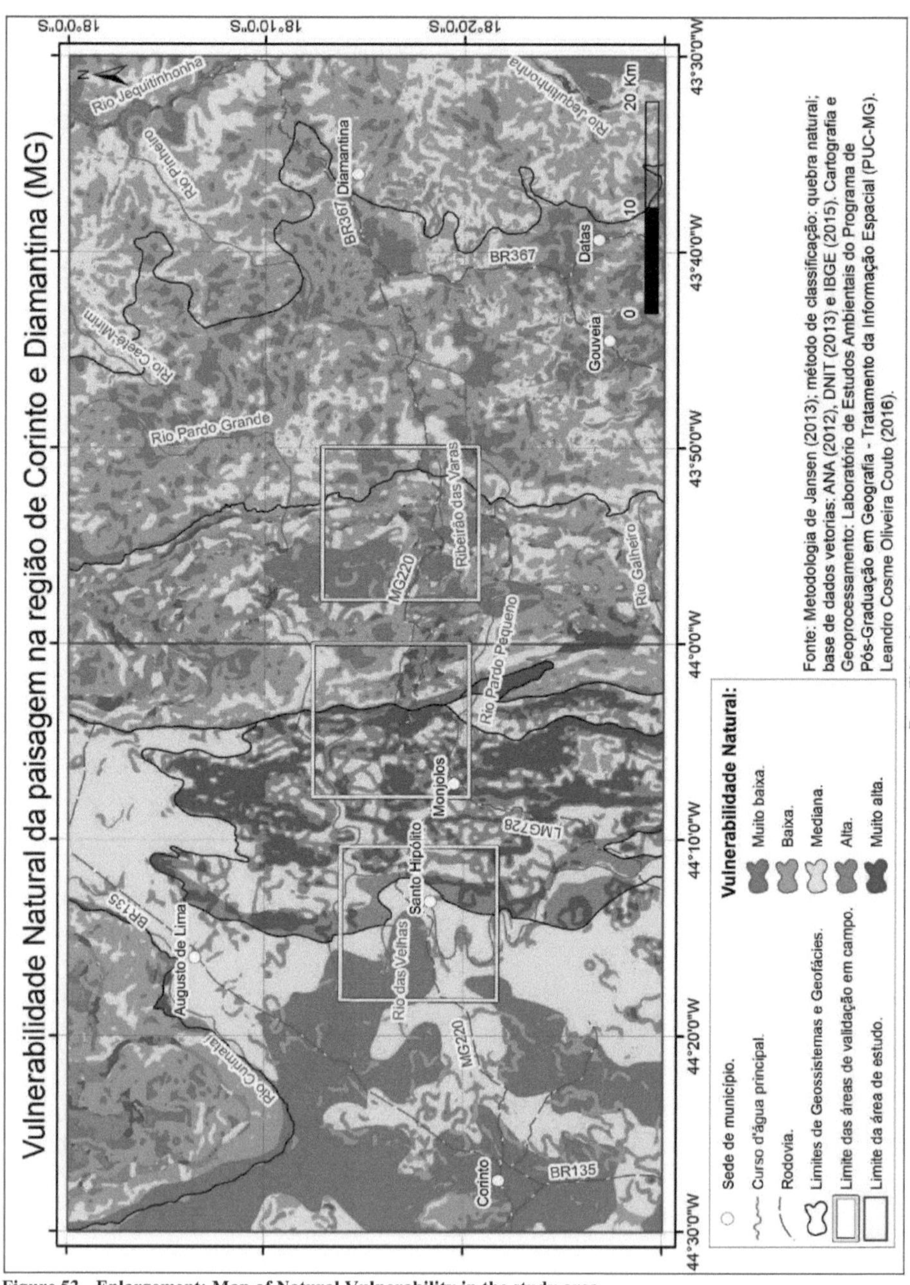

Figure 53 - Enlargement: Map of Natural Vulnerability in the study area.

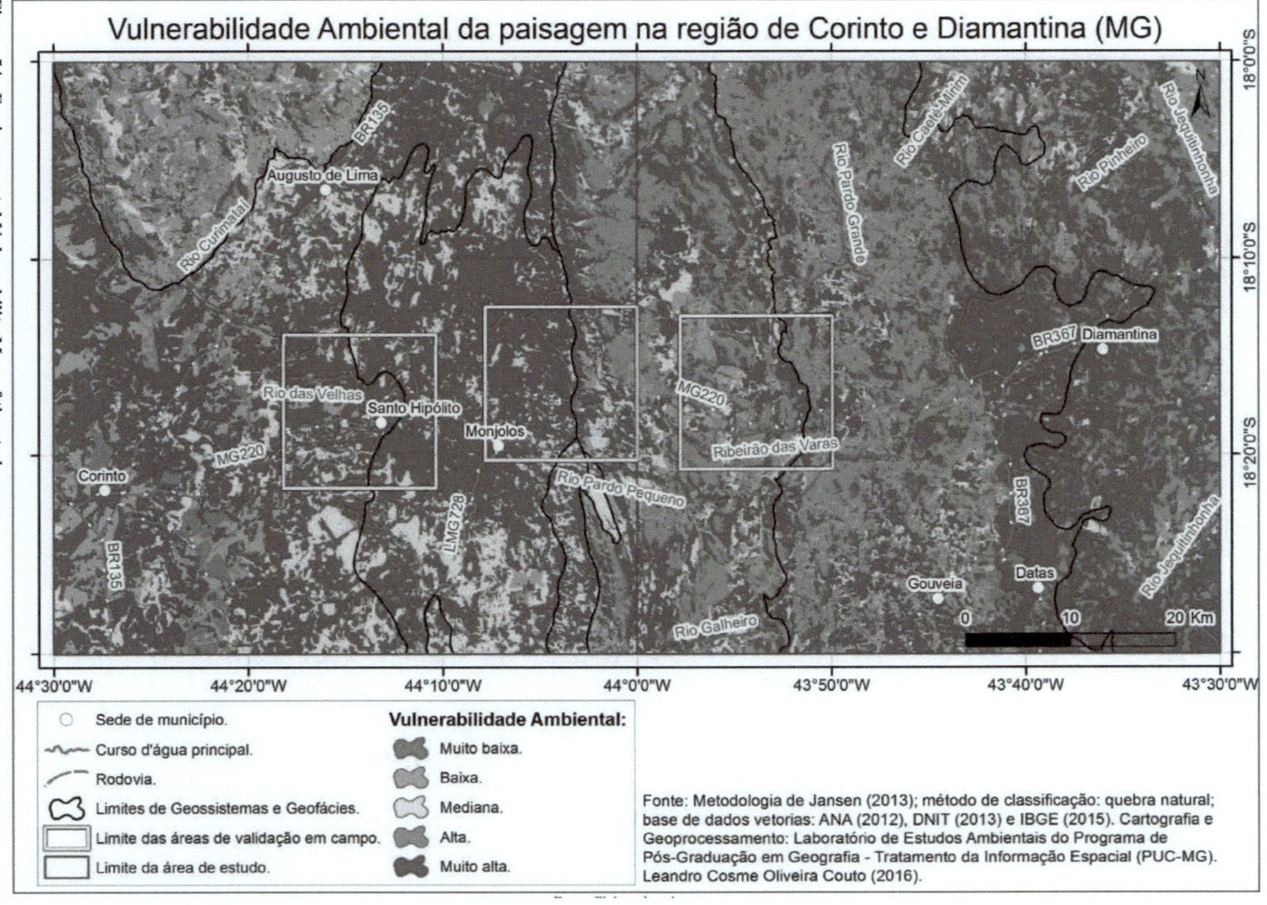

Figure 54 - Environmental Vulnerability Map of the study area.

Printed by Books on Demand GmbH, Norderstedt / Germany